Geraniums and Pelargoniums

Geraniums
and
Pelargoniums

The Complete Guide to
Cultivation, Propagation and Exhibition

Jan Taylor

The Crowood Press

First published in 1988 by
The Crowood Press
Ramsbury, Marlborough,
Wiltshire SN8 2HE

Reprinted 1989

British Library Cataloguing in Publication Data

Taylor, Jan
Geraniums and pelargoniums:
the complete guide to cultivation, propagation
and exhibition.
1. Geraniums 2. Pelargoniums
I. Title
635.9'33216 SB413.G35

ISBN 1 85223 034 7

Picture Credits

Line illustrations by Janet Sparrow
Colour photography by Frank and Jan Taylor
Jacket photographs by John Baker

Typeset by Q Set, Hucclecote, Gloucester
Printed in Great Britain by
Billing & Sons Ltd., Worcester

Contents

Foreword

There are a great many books on the family Geraniaceae, there are also a number of books which purport to deal with the geranium when in reality they are dealing with the pelargonium. In consequence there is always room for a new book which makes every effort to explain the differences and to educate the public into this most fascinating of families.

Few people who have gardens have not grown one or other of this family – they are useful as garden plants as well as being the most popular of all bedding plants used to create colour in and around our homes. Pelargoniums, however, are used in the home almost as much as outside in the garden – they provide excellent house plants which will tolerate a wide range of conditions. Colour is not the only reason these plants are so popular as many of them have aromatic foliage – even without colour they make beautifully shaped pot plants which sit nicely on a windowsill.

This book has been planned to provide as much information as possible for the skilled as well as the enthusiastic gardener, it is written in such a way that it is readable without being heavy. It is possible to discover the many different forms of this family and from this information decide which of the groups you would be interested in cultivating. All forms, from the zonals to the ivy-leaved and then into the species, are covered with a wealth of information to assist in fairly accurate identification.

Flower colour and form are well described as well as accurate explanations of growth habit. Leaves and stems are often the reason why many plants are grown and this book should enable the reader to derive pleasure and interest as well as a desire to grow them.

I recommend this book to the enthusiast and amateur alike.

R.A. Stephenson MVO

Geraniaceae Family Tree

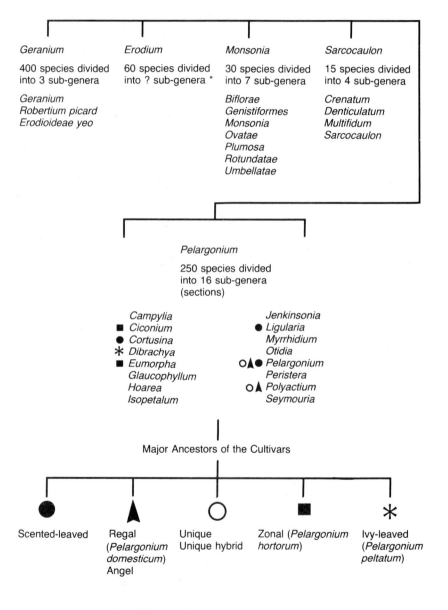

Geranium

400 species divided
into 3 sub-genera

Geranium
Robertium picard
Erodioideae yeo

Erodium

60 species divided
into ? sub-genera *

Monsonia

30 species divided
into 7 sub-genera

Biflorae
Genistiformes
Monsonia
Ovatae
Plumosa
Rotundatae
Umbellatae

Sarcocaulon

15 species divided
into 4 sub-genera

Crenatum
Denticulatum
Multifidum
Sarcocaulon

Pelargonium

250 species divided
into 16 sub-genera
(sections)

Campylia
■ Ciconium
● Cortusina
✳ Dibrachya
■ Eumorpha
Glaucophyllum
Hoarea
Isopetalum

Jenkinsonia
● Ligularia
Myrrhidium
Otidia
○▲● Pelargonium
Peristera
○▲ Polyactium
Seymouria

Major Ancestors of the Cultivars

Scented-leaved

Regal
(Pelargonium
domesticum)
Angel

Unique
Unique hybrid

Zonal (Pelargonium
hortorum)

Ivy-leaved
(Pelargonium
peltatum)

*   Accurate information difficult to obtain.

Introduction

There must be very few people who have never heard of the geranium. Although the plant has, at times, fallen out of favour, it has always leapt back into popularity charts with great gusto. Today, the plant is enjoying a deservedly prestigious come-back, rivalled only by the attention lavished upon it during the Victorian Era. Regaining its place as one of the most popular specimens grown by all who grow plants, the geranium is admired on the windowsill, in the beds and borders of the garden, and in showmen's collections in greenhouses or conservatories.

The range of types, sizes, colours of foliage and blooms, scents and habits is endless. These diverse attributes, and the reasonable ease of its cultivation, make the geranium the subject of a fascinating hobby, love or even addiction for many people. However, how many people are aware of the other plants in this family and how many are aware of the difference between a geranium and a *Geranium*?

To find out how the confusion between pelargoniums and geraniums began we have to go back some 200 years to look at the introduction of the system of botanical naming of plants. Although botanical Latin can be very confusing to the layman, it is an international language used to name plants, using either a description of the plant's origin, its form, its habit, its discoverer, or any other notable feature, and can be very valuable. Every species of plant is known by a name consisting of two words – the binominal. The first word is common to all species within a group of closely related species – the genus – and is written with a capital letter, e.g. *Geranium*. The second word is unique to the species, e.g. *Geranium molle*. These are Latin words and it is conventional to print them in italic or, if hand-written or typed, to underline them. It is acceptable to use only the first letter of the genus as a capital after the first use, if it is planned to use the genus name many times in close succession.

In those early days the plants we know today as *Geranium*, *Erodium* and *Pelargonium* were grouped into the family *Geranium*. Many new species were also being discovered,

9

Main distribution of the family Geraniaceae occurring in the wild.

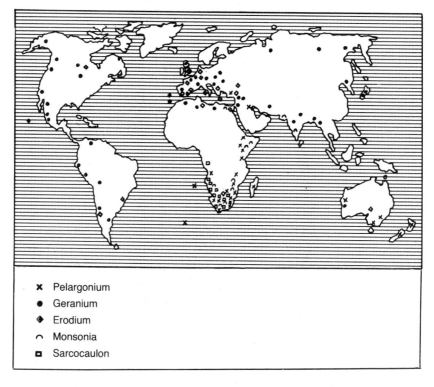

✗	Pelargonium
●	Geranium
◆	Erodium
∩	Monsonia
▫	Sarcocaulon

enlarging the family to such a haphazard degree that in 1789 a scheme to separate the family into three genera was introduced. However, although this system was accepted by botanists, it was not accepted by the public nor by many nurserymen. During the nineteenth century, hybridists developed the beginnings of the plant which most people cultivate today and use in bedding schemes, greenhouse decoration, and so on – the zonal pelargonium, or as millions prefer to call it, the 'geranium'. When the name 'geranium' appears without a capital letter this is its common name and not a botanically correct name. So, what is normally called a geranium belongs botanically to the *Pelargonium* genus of the Geraniaceae family.

To confuse the issue there are *Geranium*! To complicate it even further in the 1960s the family Geraniaceae was reorganised and now includes five genera as follows:

1. *Geranium* from the Greek, *Geranos*, meaning crane.
2. *Erodium* from the Greek, *Erodios*, a heron, hence heronsbill.

10

3. *Monsonia* named in honour of Lady Anne Monson, a great-granddaughter of Charles II and a frequent correspondent of Linnaeus.
4. *Sarcocaulon* from the Greek, *Sarcos*, meaning fleshy and *Caulon*, meaning stem.
5. *Pelargonium* from the Greek, *Pelargos*, a stork, hence storksbill.

These five all possess the same formation of seed capsule (resembling the beak of a bird) and so are classified botanically into the same family. It is about these five genera and their 'offspring' that this book has been written – for the beginner, for the enthusiast, and for those who need encouragement. The latter category, it is certain, embraces all of us at some time or other.

All noble families own a family tree. The Geraniaceae family also deserves a right to that honour. By inspecting it carefully the reader will, no doubt, be enlightened. The origins and forms of the family and the ancestry of old favourites grown, perhaps unappreciated, for generations without a second thought, will appear clear. It will most likely be a useful exercise to refer to the tree whilst reading each of the following chapters.

Commercial growers, seed merchants, parks and even botanical gardens do not help to rectify this naming confusion, although one can sympathise with their dilemma. They *do* know better but have to make a living, and public parks and gardens must cater for the general public as well as the gardener or enthusiast. If they called these geraniums by the name *Pelargonium*, a large percentage of the general public would have no idea of what was meant. But a silent revolution is under way. Many of the interested parties are noting the correct name in brackets or using the true name with the common name also mentioned. This is a major step forward and should be encouraged. It is up to us as *Pelargonium* and *Geranium* lovers to help educate the general public, and we owe it not least to the true *Geranium*, which has for too long been the poor relation in the family, to get things right! If for just one week the word geranium was made taboo in the *Pelargonium* grower's greenhouse and all visitors given a brief explanation, this would at least begin to ease the muddle. For a further week, the hobbyist could refer to his plants by their type names, for example, zonal pelargonium, ivy-leaved pelargonium, and so on.

The most likely question a visitor will ask is 'How can I

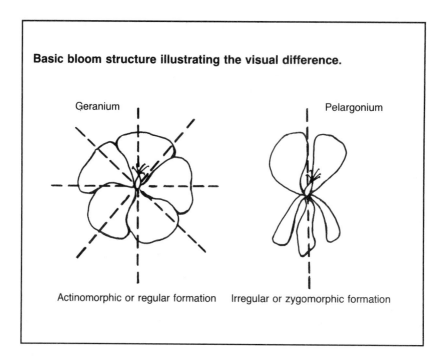

Basic bloom structure illustrating the visual difference.

Geranium

Pelargonium

Actinomorphic or regular formation   Irregular or zygomorphic formation

distinguish between the two just by looking?' The obvious indication of the difference will be the flower; botanical 'unseen' comparisons of the plant will be evaluated under the relevant sections, but to explain any tangible difference these illustrations may help.

To emphasise the flower formation of some of the modern varieties of *Pelargonium*, remove one bloom from the plant and, holding it from the rear by the flower stalk, press the bloom down on to a flat surface – usually the petals will take on the typical 'two petals uppermost and three petals below' formation.

With this basic understanding of the family and the difference between the two genera that are more usually grown we can continue to explore the five members of the clan in more detail, beginning with *Geranium*.

1

Geranium

Geranium is from the Greek, *Geranos*, meaning a crane and referring to the beak-like shape of the ovary or fruit (schizocarp) comprising five joined sections (carpels) forming a capsule. This capsule is mounted on a long, straight, tapering column called a style – in plants with a beak-like seed pod resemblance it is often called a rostrum.

When fully ripe the schizocarp splits into five separate sections (carpels) each normally containing one seed. The seed dispersal begins with each carpel breaking away from the base of the style and then dispersal is effected by shooting a short way from the parent plant still inside the carpel and with the tail (awn) attached – this is called the *Erodium* type seed dispersal.

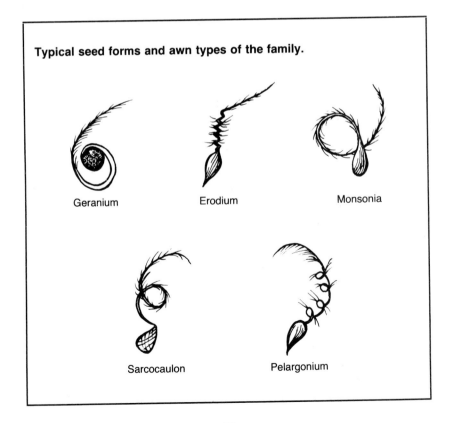

Typical seed forms and awn types of the family.

Geranium   Erodium   Monsonia

Sarcocaulon   Pelargonium

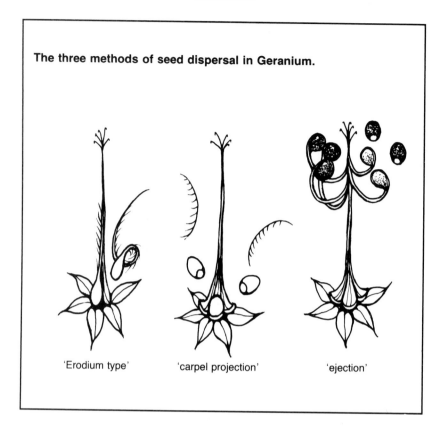

The three methods of seed dispersal in Geranium.

'Erodium type'   'carpel projection'   'ejection'

Alternatively, the carpel discharges without the awn to a short distance from the plant, dropping the awn nearby – this is called carpel seed dispersal. Ejection is the final method – the carpel complete with awn stays attached to the style and using an explosive action propels the seed from an aperture in the carpel and shoots in an upward and outward motion quite a distance from the mother plant. In all cases the style remains as a central column on the stem of the seed head. The size of seed varies enormously from less than a pin-head to about (3 mm) (0.1 in.) across, depending on the species. Each seed is dull and smooth (some do have a rib), fairly oval in shape with a medium hard coat when dry. The colour is normally mid to dark brown when ripe.

These three methods of seed dispersal determine the three sub-genera of the genus, namely:

1. *Erodioeae* (*Erodium* type seed discharge).
2. *Robertium* (carpel projection dispersal).
3. *Geranium* (ejection principle of seed dispersal).

Each sub-genus also contains sections and groups but it is deemed wise to allow the reader to assess his or her knowledge and investigate from sources of a strictly botanical nature if so desired.

The name *Geranium* was used by Dioscorides who lived at the time of Nero and the genus can boast about 400 species – some are herbaceous perennials and some are annuals. A few of these have been found in South Africa, Australia and New Zealand, and also in smaller numbers on the Islands of Hawaii, but by and large the vast majority are to be found growing in the northern hemisphere. It is the latter that are grown as hardy garden plants. The British Isles has an ideal climate for the few forms growing wild here. 'Herb Robert' (*Geranium robertianum*) and 'Dove's Foot' (*Geranium molle*) are often classed as garden weeds. The 'blue' geranium or 'Meadow Cranesbill' (*Geranium pratense*) is well known in the hedgerows. *Geranium sylvaticum* and *Geranium sanguineum* are also British natives but perhaps not so well known and certainly less common in the wild.

Geranium macrorrhizum was one of the first mentioned in the sixteenth century and today many species, natural hybrids and cultivars are available. The origin of the *Geranium* goes back centuries and so do the stories and reports of the *Geranium* having powers to heal and quell. In fact the medieval apothecary used the stems, leaves and roots, either cooked for concoctions to allay afflictions of the kidney, genital organs, lungs and eyes and to stem bleeding, or the leaves were taken straight from the plant to be laid over bruises, ruptures and rashes to ease inflammation.

Today *Geranium* are grown exclusively as garden plants and because most can claim winter hardiness are a valuable addition to any garden. It is quite easy to purchase the more well-known species and varieties for a reasonable sum. The more unusual ones may have to be tracked down at specialist nurseries and the rarer ones might have to be grown from seed – this is often available with membership to specialist societies.

To begin a brief, general description of the *Geranium* let us start literally at the roots.

ROOTS, FOLIAGE AND BLOOMS

There are three basic root formations. Fibrous roots are the most common type of root in *Geranium* and as they are also the most common type of root in the plant kingdom require no further explanation, except that the top growth from the rootstock can either be stout and upright or lax and scrambling. The second type are tap roots, similar in structure to the carrot with small roots along the centre rootstock which sometimes may be found at the soil's surface. These take in and store water and plant food in times of plenty to restore the plant when nature creates famine. Usually in this type the top vegetation grows in the form of a crown on fairly short stems making a rosette. The third possess tubers, similar to the dahlia, of varying size and shape sometimes connected with a thread-like root giving an appearance of a string of beads. The tubers may be known as stolons or rhizomes and are used for food storage and some for the production of new plants.

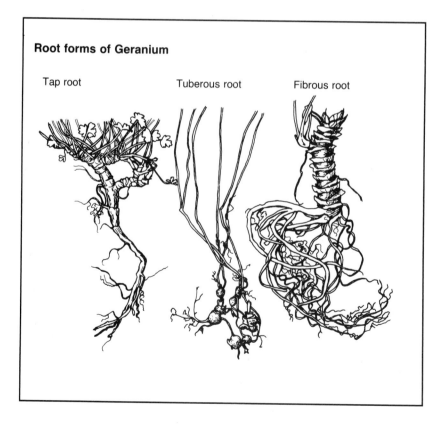

Root forms of Geranium

Tap root   Tuberous root   Fibrous root

Often this type of root structure will wander a considerable distance through the soil to seek out ideal food supplies and soil consistency. It has been known for these creeping rhizomes to remain underground for one or two years and fail to make any top growth during this period.

Foliage is extremely variable in size, from 0.5in. (1 cm) to upwards of 20 in. (50 cm). The shape may be round or palmately divided at the base resembling the common sycamore (*Acer pseudoplatanus*). All, to some degree, have divisions. These divisions are either shallow or so deeply cut the leaves resemble those of the 'Spring Adonis' (*Adonis vernalis*). Down or short hairs are evident giving some species an overall grey tone to the normally mid to dark green leaf colour; a few varieties produce russet hues in the foliage intensifying in the autumn; many are totally evergreen and variegated foliage can be found in one or two forms. At the junction of the leaf stalk (petiole) and stem, pairs of stipules persist.

Blooms are often saucer-shaped, regular in formation, with five petals (there are some cultivated double forms) all of which are uniform – sometimes notched at the tip, sometimes reflexed or flat – and arranged radiating symmetrically. The flowers contain ten stamens, all fertile or possessing pollen. They are usually displayed in pairs on stems rising above the foliage. Pink, purple, magenta, mauve, blue and white, often with speckles or striations, compile the great colour range of the flowers (inflorescence). In fact it is only the true scarlet red portion of this section of the colour spectrum that is lacking. The Royal Horticultural Society's Colour Chart (1986) Fan No. 2 contains the complete colour range of the hardy *Geranium*.

In stature the range is similarly comprehensive, the smallest growing only 1 in. (2.5 cm) high and the tallest to 4 or 5 ft (125cm to 150cm) high with differing circumference. Being generally tolerant of sun and shade, and often fairly deep shade, members of the *Geranium* genus prove exceptionally adaptable plants for the garden and with their ground cover abilities emphasise their usefulness in the modern, low maintainance garden – either as a bed complemented with other subjects such as *Hellebore* and *Anemone japonica* or under trees and shrubs where they may scramble around and between to their hearts' content.

CULTIVATION

Herbaceous borders, alpine gardens, rockeries or wild gardens are the usual habitat for *Geranium* 'in captivity'. Most soil types are suitable although if heavy clay is present digging in gravel, sharp sand or a mixture of both will help root penetration and ensure good drainage which is particularly important for those with dense crowns. Rotted compost or farmyard manure dug into a new bed or into an existing bed in the autumn will form a basis for good growth and later in the year a general, preferably organic, feed will be beneficial. However, take care not to over-feed and encourage too much soft, leafy growth as soft growth will not resist disease.

Acidity or alkaninity (pH factor) does not appear to be crucial but familiarisation of the natural habitats of the species in question will enrich the grower's success in acquiring perfectly grown specimens.

Staking or supporting some varieties may be necessary although siting the plants known to be lanky growers near shrubs through which they may grow and be supported, will eliminate the need to use canes or stakes. For those needing gentle support, a few twigs inserted carefully at the beginning of the growing season is all that is required but it may be that the twigs will have to be heightened as the plant grows. The type of wire plant support comprising a circle with an inner trellis of wire can be purchased readily – these are ideal and can be adjusted to keep pace with growth.

It is important to remove any spent blooms throughout the season and to take off dead or dying foliage. Both these actions will encourage new growth and extend the flowering season as well as deter pests and diseases associated with decaying vegetation.

As mentioned earlier, *Geranium* will be happy in sun, part shade or sometimes in heavy shade. Exposed situations do not seem to be a problem to most kinds. Although some have a fairly short flowering period, others will begin blooming in May and continue through until the first hard frosts of winter. A succession of flowers and attractive foliage can be achieved for about ten months of the year. It is advisable to house those few varieties that are not reliably hardy in Great Britain in a cool greenhouse for the winter months.

PROPAGATION

Seeds

Whenever practical seed should be sown as soon as possible after gathering thus ensuring freshness and an ideal outer skin condition. Purchased or old seed will benefit if the seed is either nicked shallowly or a groove made with a nail file to allow better water penetration. Many varieties of *Geranium* will germinate readily with conventional methods and, if seed is sown by mid–spring, will often bloom in the late summer of the same year. If problems regarding germination are expected it may be helpful to expose the seed to frost. This is called stratification and is the procedure followed by seeds of many types of plants from the temperate regions, which quite naturally use a process to chill the seed and break its dormancy. The simplest method to copy this is to sow seeds in a pot or tray containing good quality seed compost, label the batch and leave outside, away from scratching animals and covered with wire to protect from birds. The seeds will then experience the winter elements and by late spring should be shooting.

There is, however, a short cut. By fooling the seed into thinking winter has passed it is possible to encourage germination in a shorter time. It must be mentioned that as the resulting seedling will probably be growing during an unsuitable time, care will have to be taken in keeping the seedling happy in a mild climate until early summer when it can be planted out into the garden. If this method is to be tried, mix together a handful of sieved or rubbed sphagnum moss peat with two tablespoons of warm water in a plastic bag so that when squeezed the mix will mould. Add to this two tablespoons of horticultural sand and the seeds. This amount of mixture should be enough for thirty seeds. Label and tie the bag with a twist-tie and place in a warm place, such as an airing cupboard, for three days, then put into the coldest part of the refrigerator (not freezer) for three weeks if fresh seed or four to five weeks if old seed or purchased seed. Shake the bag each week to aerate the mass. After the chilling period sow the seed into pots or trays containing seed compost by shaking the bag of mixture onto the surface of the pot and spreading evenly. Do not cover with lids or plastic bags. Place under the bench or in a shady place. Leave to sprout, watering only when essential using a fine mist spray. For sowing seeds not stratified or those freshly harvested sow as has already been

19

explained. Germination can take from a few days to a few months.

When the seedlings first emerge, two rounded, pale green leaves are present. These are the seed leaves or cotyledons. It is at this stage when problems with damping off occur and so a gentle spray with a warm fungicide solution is recommended. More light can be provided at this stage but never strong sunlight. After a few more days two true leaves will be evident and when these are large enough to handle, carefully lever the seedling from the pot, holding it only by the foliage and replant carefully and neatly into a tray of John Innes No. 1 potting compost or equivalent. Allow the seedlings to grow on into small plants which can then be planted in the garden when large enough.

Division

Splitting and dividing of clumps is the easiest method and supplies the quickest return for one's labours but not all types can be multiplied in this way. Division is best carried out from early December to March, avoiding periods when the ground is frozen or frosts and icy conditions are threatened. To prise the clump apart, sink two hand or garden forks back to back into the soil at the centre of the plant and with a contrary levering motion gently part the plant in two or more pieces. This method will tease out the roots and is preferred to cutting up pieces with a spade. Replant the clumps where desired after removing any dead or damaged vegetation. Tubers or rhizomes may also be eased out of the ground using this method, potted up and left to grow and shoot.

When the plant is too small or it is not practical to split the stock, cuttings may be the ideal way of ensuring a new plant. Cuttings can be taken from both stem and root.

Cuttings

Begin by assembling all the required equipment, namely a clean, razor-sharp knife, pots or trays in which the cuttings are to be housed, labels and label-marker (make sure this is waterproof), a reservoir large enough to accommodate the pots or trays, a watering-can with a fine rose or a sprayer and ample compost. The compost should be capable of retaining moisture and at the same time capable of free drainage, so a 50 per cent mix of either horticultural sand, perlite or vermiculite and either sphagnum moss peat or proprietary seed compost is preferable. The chosen two ingredients should be well rubbed and aerated, as in the method used to make pastry! Fill the pots or trays to the top with this mix and press down lightly with a specially made block or the base of a clean pot. Into the reservoir pour warm water (warm water will penetrate the compost more quickly). If desired a fungicide may be added to the water.

There are some types of peat pellets or peat pots on the market which are specially produced for the rooting of cuttings. These are expensive but, if only one or two cuttings are to be taken, are perhaps more suitable especially in the situation of limited garden or growing space. It is extremely important to take care with the watering of these pellets or pots. Their main advantage is that one can see when the cutting has rooted because the roots appear through the side wall or net covering. Problems may occur at the planting on stage because the pellets or pots are mainly peat and the resulting rooted cutting is to be planted in the garden, which is unlikely to be very peaty. Therefore, the young plant may require a weaning programme onto a loam or garden soil 'diet'.

Allow the pots or trays to soak in the reservoir until moisture can be seen breaking the compost surface and leave to drain for at least half an hour. If the peat pellets are to be used place these in the same solution until they have become well wetted and risen to their full size. Drain well. For peat pots the same solution can be used – lay the pots in the water until well soaked then fill with the compost and soak again until the surface breaks with the water. Leave to drain. It is essential that peat pots are soaked before filling or the compost will dry out immediately as the outer peat wall absorbs the moisture.

Filling the pots and trays first allows time for them to drain whilst the cuttings are being prepared. Choose a semi-mature, preferably non-flowering shoot from a healthy plant. Cut off just below a leaf joint (node). Neaten the stock plant to just above the leaf joint and dust the wound with a fungicide powder (Captan) or cover the wound with dry horticultural sand if there is evidence of major damage to the plant. Trim the cutting making the length about 2.5 – 3 in. (6 – 8 cm) and cut straight across the stem just on or just below a leaf node. Either a cutting from the growing tip or an internodal portion may be used. Take off the lower leaves, any blooms or buds, stipules and any damaged parts. Write a label with the name and any other relevant data such as date, mixture used, etc. Carefully handling the cutting by the foliage where possible, insert the cutting into the compost to a depth of about 1 in. (2.5 cm). Tap the tray gently on a flat surface to settle the compost around the end of the cutting – this is important because any air pockets at the base mean that the compost is not in contact with the cutting and so it will not be able to root into its surroundings. Do not cover and for the first few days put the pots or trays under a bench or in the shade to recover; afterwards leave in a cool light place (but out of direct sun) if the cuttings are taken during the late spring or summer, if taken later a cool greenhouse will be an advantage and possibly some bottom heat of about 60°F (15°C). Water only when needed using warm water, cold water can be a shock to the system, both for plants and humans! Normally the cuttings will begin to root from twenty days depending on varieties. If the cuttings appear healthy and sprightly, and are taking more water it is possible they have begun to root and a very careful examination of the compost is the only way a beginner can be sure of this. Later, experience will give the answer. When they have rooted well, plant in the final garden position watering in adequately and repeat watering when necessary in periods of drought until the young plant is established – then let nature take over. From May to September is the ideal period to take stem cuttings as a trial, when the plants have begun to break their winter dormancy. Some species and cultivars do not supply top vegetation suitable in quantity or size for stem cuttings and so cuttings from roots are possibly the solution.

Root Cuttings

These are treated in a different way and are a useful alternative of propagating when a specimen will not provide material for other methods of propagation or when the plant could be spoilt if divided or cut or perhaps will not yield seed for one reason or another.

Assuming the plant to be muliplied is *in situ*, carefully, with a hand fork, remove a section of soil from one area near the plant's rootstock until roots are visible. Using a clean sharp knife, cut sufficient roots from the rootstock as near to the plant as possible. Replace with soil to which has been added horticultural sharp sand. Do not water for a day or two to allow the wounds to dry off and the healing process to begin.

With the root pieces, clean off, between finger and thumb, any surplus soil, trim off any small or damaged ends then cut into lengths of 2 – 3 in. (5 – 8 cm). Place these into a small plastic bag to which a fungicide powder such as Captan has been added and shake the bag to cover the roots with the powder. Fill a pot or seed tray with the cutting mixture as previously mentioned but only just moistened. Push in the root pieces until 0.5 in. (1 cm) is left showing and cover to the top of the pieces with dry horticultural sand. Do not water. Place in a heated propagator at 65°F (18°C) or in a cold frame, or in a cool greenhouse. If heat is provided the root buds should become evident from four weeks. Leave the cuttings in the pot or tray until the roots have filled the soil. During the rooting, keep watering to an absolute minimum until the buds have developed. Later in the top-growth cycle a weak feed of liquid general fertiliser may be given. Plant out into the open site or bed when foliage is strong and plentiful.

The plant's natural dormancy period, generally in the autumn, is the best time to use this method of propagation but it is advisable to experiment as species or varieties may vary in their adaptability to this system. This method cannot be used on tap-rooted or annual types and seed is the best way to reproduce in these cases.

Layering

This is another form of production acceptable quite readily to some *Geranium*. Where plants send out long leafy stems covering the ground like runners decide where on the stem to take the cutting and fork into the soil at that point a quantity of horticultural sand. Just below a leaf node and on the underside of the stem, make a small, shallow incision with a clean sharp knife taking care not to cut right through the stem. Lay the stem onto the prepared soil and hold in place, over the cut, with a wire staple or hair-pin. When rooting is obvious, and this may take from three weeks to a few months, cut the stem from the mother plant, carefully dig up the rooted portion trimming off unnecessary leaves and stem pieces and plant in its new position. As this procedure will normally be carried out in the open there will not be any need to water except in times of drought. The months between June and September will prove to be the most likely time for success using this method.

HYBRIDISING

For amateurs the crossing of *Geranium* may prove a little exasperating mainly due to the plant usually being grown in an open situation and not in a controlled environment. This, of course, leaves the flower accessible to insects, birds and even the gentlest breeze, all of which provide the minimal amount of movement needed to tranfer the pollen onto the stigma so creating a chance crossing. Serious hybridising can be carried out in the open garden if a careful watch for maturity in the anthers can enable them to be removed at the correct time so that the flower will not fertilise itself. A greaseproof paper bag, which will be fairly waterproof, is placed over the bloom for one day and the flower is then inspected for receptiveness of the stigma. At this time pollen from a chosen flower is dusted onto the stigma, the bag replaced and left until the seeds are formed and afterwards ready for dispersal. This is a crucial period and careful inspection is necessary to time the moment when the seeds are at point of despatch or all the seeds could be lost. Experience and familiarity of the seed setting cycle will help to ensure success in catching the ripe seeds on time.

Very occasionally a lovely new variety is perfected from a chance crossing as well as calculated efforts but patience is needed to this end and only worthy crosses should be retained, or any that are suitable as a basis for a more comprehensive breeding programme. It is extremely vital that accurate and detailed written records of any hybridisation programme are kept up to date. If the cross happens to be very different from any other form and so meritorious then it is worth studying a copy of the *International Code of Nomenclature for Cultivated Plants* and a copy of the *International Code of Botanical Nomenclature* – both these are available from the Royal Horticultural Society in London. The naming code you must adopt depends whether your new plant is a cross between two species or from a species and a cultivar or from two cultivars.

Remember there are many hundreds of species and cultivars so do check that the cross is really something special before ideas of flooding the market develop!

PESTS, DISEASES
AND DISORDERS

Fortunately these are fairly few and they rarely cause great problems.

The main area for concern is at the propagating stage and this is usually eradicated if hygiene and common sense prevail. For example, wash pots, trays and bench tops regularly with Jeyes Fluid or similar following the manufacturer's instructions. Sterilise knives with methylated spirits, Jeyes Fluid solution, etc. wiping dry with kitchen paper or tissues. Always use fresh, good quality compost of the right formula for the job, either purchased, in which case it should have been sterilised, or one's own mix adequately sterilised. Dipping leafy cuttings in a fungicide solution will minimise fungal diseases, damping off, and so on. Use clean tap water (warm) unless it is perfectly certain that only clean rain water has been collected.

Mature plants may also suffer from bacterial or fungal problems due to spores or bacteria entering wounds or plant debris. Use a fungicide powder to combat this.

Occasionally *Geranium* are struck by a form of rust which is noticeable at times as yellow spots or blotches on the upper surface of the leaves with a rusty brown powdery deposit on the underside. Take off affected foliage and destroy totally, then spray with a fungicide such as copper sulphate or Benlate. Ensure that the clump of the plant is not too dense; if so prune away some of the centre growth and this will enable air to circulate more freely through the clump.

Caterpillars are sometimes a menace. Either pick them off and drown them in a bucket of water, after making certain that they are not a rare variety or, if that is too large a task, spray with a proprietary pesticide formulated for caterpillars and weevils, etc.

Slugs and snails are also a problem, particularly in a late spring just when the new growth appears at the same time as these pests become more active. Again, unless time is available for a midnight slug hunt or saucers of beer traps are laid, purchased slug and snail killers may be the only solution. Rotting vegetation, areas of nearby rubble and damp, dark, cool places are a haven for slugs and snails and so avoiding these would deter any major destruction.

Vine weevils and crane-fly larvae (leather jackets) attack plants at soil level or just below. The difficulty is that often their presence goes unnoticed until the plant collapses or wilts due to being eaten through at the crown or main root. If pests are suspected, carefully dig out the plant and inspect. If the damage is not total, remove and kill the grubs. Wash the roots in an insecticide liquid. Replant in a new position, watering the original site with the same insecticide.

Infestation of aphids and other flying pests can be controlled with proprietary insecticides if the affliction is serious. Otherwise, let the finger and thumb do the work of destroying them, or allow 'friendly' insects and birds the chance of a feast to try and keep pests under control.

Only use pesticides and fungicides as a last resort. Prevention is much better than cure and a much safer course to take. Nature will take care of many problems that appear in the modern garden.

Disorders, often due to adverse weather conditions, such as frost damage or wind scorch, are sometimes wrongly diagnosed as a disease, as is disfiguration due to careless weed control chemicals or insecticides. Always look for the simple explanation before embarking on the task of eradicating pests and diseases – simple removal of foliage is perhaps all that is required. Deficiency disorders are more of a problem for the beginner, although not common with plants grown in the open garden. There are many helpful leaflets available from fertiliser manufacturing companies with coloured plates showing examples of nutrient deficiencies in plants to assist in detecting which element is lacking in the plant's food supply.

A piece of foliage or plant shown to an expert grower will usually lead to a knowledgeable solution, but please put the piece of vegetation in a sealed plastic bag, just in case the plant has some contagious or infectious disease.

Remember, healthy, well-grown plants, healthy soil and a lively atmosphere will keep the majority of plant problems at bay in the garden.

EXHIBITING

Some specialist shows do hold classes for *Geranium* ('Cranesbills'). Due to the size of most of the specimens and the fact that *Geranium* do not adapt too happily to pot culture, exhibiting or showing is limited. Some exhibitors favour the procedure of potting up plants only a few weeks before a show. Others rely on growing the more diminutive of the

Planting plan for a geranium bed.

Rear fence – wall – screen – hedge etc.

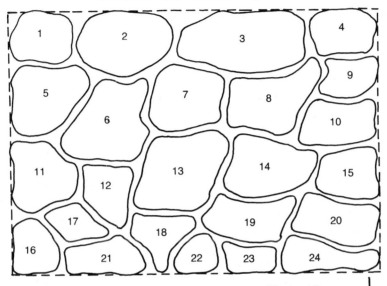

Interplanting of Hellebores and Anemone Japonica would extend the flowering period.

↓
N

Initially three of each species of variety is recommended.

Key

Geranium

1	sylvaticum wanneri 'Briquet'
2	psilostemon
3	pratense
4	wlassovianum
5	grandiflorum plenum
6	phaeum *var.* lividum
7	sylvaticum 'Album'
8	phaeum
9	ibericum
10	endressi 'Rose Clair'
11	'Russell Pritchard'
12	rectum 'Album'
13	macrorrhizum in variety
14	versicolor
15	macrorrhizum variegatum
16	dalmaticum
17	renardii
18	orientalitibeticum
19	nodosum
20	sanguineum and sanguineum 'Album'
21	sanguineum lancastriense
22	rivulare
23	cinerium 'Ballerina'
24	sessiliflorum in variety

genus or the ones which are easier to care for in pots and pans and are simpler to transport to shows.

Success depends on the cleanliness of the plant and its container, the quality, shape, quantity and colour of both foliage and flowers and a general apparent well-being of the exhibit.

Geranium are not the simplest of plants to grow for showing but certainly well worth a try, even if only to hear the congratulatory comments from those who realise the degree of difficulty that has been experienced by the exhibitor as well as comments from the general public, many of whom will be heard to say that they have never seen a plant like that before!

Floral Art classes are sometimes staged, asking for arrangements containing *Geranium* blooms. These can be very charming especially when used with their own foliage. The blooms will last for more than a week in water and it has been known for 'the other half' to excel in these classes whilst the 'partner' exhibits in pot classes. Many flower arrangers have become avid growers of these lovely plants and vice-versa.

PLANT LISTS

The plant lists throughout this book are only a fraction of the varieties available and are just suggestions, most of which the author has grown. Some are suitable for the beginner and some for the person who has already grown members of the Geraniaceae in one form or another and wishes to try one or two of the more unusual and challenging ones. Brief descriptions, observations and, when necessary, explanations are given.

There are many nurseries specialising in the plants and seeds mentioned in this book who produce excellent catalogues for a few pence plus postage – these all list so many that to duplicate here is unnecessary. Going through the catalogues from specialist nurseries and seed firms will provide the reader with adequate choices, interesting reading and descriptions of plants available for purchase.

Following each description are codes as follows: B – suitable for a beginner; E – suitable for an enthusiast.

GERANIUM PLANT LIST

G. *cinerium* 'Ballerina': Cut and lobed, round, grey-green leaves. Lilac-pink flowers with dark purple veinings. Suitable for the rock garden. 4 in. (10 cm).B

G. *clarkei* 'Kashmir White' (syn. G. *rectum album*): Leaves deeply divided. Flowers white with pale mauve veins. 10–12 in. (25–30 cm).B

G. *dalmaticum*: Palmately lobed leaves of mid-green forming a low cushion plant. Leaves tinted red in autumn. Blooms of shell pink. There is a white form. Rock garden in sun or light shade. 6 in. (15 cm).B

G. *endressii* 'A T Johnson': Mid-green palmate and lobed leaves. Silver-pink blooms. Good in shade. 12 in. (30 cm).B

G. *Hymalayense plenum* 'Birch Double': Dwarf and vigorous. Flowers lavender blue. 10 in. (25 cm).E

G. *ibericum*: Five to seven lobed mid-green leaves. Blooms of violet blue with feathered purple veins. 18–20 in. (45–50 cm).B

G. *maderense*: Large mid-green deeply divided and lobed on brownish leaf-stalks. Flowers on top of plant and of magenta-pink. Needs greenhouse winter protection. 60 in. (1.5 m).E

G. *macrorrhizum*: Five-lobed mid-green leaves. Autumn colour with evergreen, aromatic foliage. Pale magenta-pink flowers. There are many varieties including a variegated form. Good for shade. 12 in. (30 cm).B

G. *nodosum*: Three or five lanceolated toothed leaves, bright green. Swellings above leaf nodes. Flowers funnel-shaped of purplish-pink. Good for shade. 10–18 in. (25–45 cm).B

G. *orientalitibeticum*: Underground tubers, leaves deeply cut and with marbling of light greenish-cream on mid-green foliage. Flowers, pale magenta-pink with large white eye. Dwarf and trailing. Suitable for rock garden, may have to be kept under control.B

G. *phaeum*: 'The Mourning Widow': Large leaves deeply cut with maroon blotches at times. Darkest purple-maroon flowers. There are pale and white forms. 15–24 in. (38–60 cm).B

G. *pratense*: 'The Meadow Cranesbill': Leaves of mid-green, five to seven lobed, deeply divided on long petioles. Flowers blue. There are white, grey and double forms. Good for the wild garden, enjoys growing in grass and some shade. 12–24 in. (30–60 cm).B

G. *pratense plenum* 'Album': As above but not so tall and vigorous. White double blooms with greenish centre at times.E

G. *psilostemon* (syn. G. *armenum*): Mid-green five lobed leaves, palmate. Flowers slightly cup shaped and magenta with large almost black eye. Will tolerate some dry shade. 3 ft (90 cm). B

G. renardii: Hummock of leathery, grey-green leaves well veined. Autumn colour of foliage. Flowers off white with definite purple veins. Good for shade at front of border. 12 in. (30 cm). B

G. sanguineum lancastriense (syn. *G. sanguineum striatum*): Deeply lobed and cut dark green leaves with autumn colour. Flowers saucer-shaped and flesh pink with pale veinings. Flower bearing shoots tend to scramble. 4 in. (10 cm) in height. B

G. sessiliflorum: Rounded and divided foliage colour ranging from brownish-green to dark brown. Flowers tiny and white. Good for front of border and rock garden. 2–3 in. (5–7 cm). B

G. sylvaticum 'Album': Rounded seven lobed greyish-green leaves. Flowers white. 24 in. (60 cm). B

G. traversii: Silver foliage of round divided and evergreen habit. Flowers pink and quite large for plant size. Needs greenhouse protection or winter covering outdoors. 7 in. (15 cm). E

2

Erodium

The name *Erodium* comes from the Greek word *Erodios* – a heron – so 'Heronsbill' is one of the common names for this genus. Confusingly another common name often used is 'Storksbill'. Again the name is describing the shape of the fruit which contains five seeds. In shape the seeds are in keeping with the rest of the family with the colour of the husk ranging from beige and light brown to rust. The *Erodium* has only one method of seed dispersal in that the carpels separate from the base of the rostrum or style and curl upwards, the awn spiralling. Then the seed in its casing (mericarp) is detached with some force, springing a distance from the parent plant. The awn is equipped with short bristles and spirals – the intensity depending on the humidity. Generally more seeds are detached and sprung further during low humidity. The seeds are screwed into the ground by this spiralled awn to await ideal conditions in the soil for germination.

This genus contains roughly ninety species, some annual, some perennial, some technically shrubs or sub-shrubs. In the British Isles there is one species growing naturally as a wild plant, *Erodium cicutarium*. It inhabits fields, wastelands and waysides so prolifically no wonder it is known as 'Common Heronsbill' or 'Common Storksbill'. As it relishes weedy places and plenty of nitrogen in the soil it may often be found in cultivated fields and gardens or at the path edge of such places and looked upon as a weed. *Cicutarium* is derived from a classic word meaning hemlock, to which the leaves are similar in shape. Hemlock is the poisonous herb often used during the middle ages as part of a mixture to kill pain!

Most *Erodium* are to be found in the wild throughout Asia Minor, North America and countries north and south of the Mediterranean and Europe. Few may be located in Australia and Southern Africa. Their haunts in the wild give a clue to the differing preferences of climate and soils required in the garden. Many are winter hardy in Great Britain but some do require shelter from our cold rainy spells in winter and a few are happy to be situated in a cold greenhouse or alpine house. Usually the easy-to-grow varieties and species are available

31

from specialists in alpine or perennial plants while the less commonly grown types will have to be tracked down or grown from seed acquired from specialist societies or individuals who enjoy growing *Erodium* because of the challenge some create as well as the plant's absolute charm. Because of this plants will generally be more pricey than the run-of-the-mill perennials.

On the whole the plants are neat, compact, short and close-textured but there are one or two much larger and more open.

ROOTS, FOLIAGE AND BLOOMS

Roots

Most possess a fibrous rootstock comprising very fine and fairly short roots, in a few species the root thickens and occasionally partially rises to the soil surface looking like the rhizomes of the 'Flag Iris'. The finer rooted species are sometimes to be found in areas where the soil is sparse and the small rocks create an insecure anchorage and so the roots have the power to seek a holding in the minutest cranny – some even work their way into porous rock such as tufa.

The general habitat of *Erodium* is one of free drainage and shelter from excessive moisture. The rootstock is capable of adapting to these natural climatic conditions, taking in water when it is available and allowing the plant to rest when moisture is scarce. In such cases the food supply is also limited and *Erodium* naturally do not require vast quantities of rich nutrients.

Foliage

This is very variable, some species having extremely divided, long, tapering leaves, finely cut into lobes and resembling the foliage of the carrot. A few others have bold, lobed leaves. All are covered with woolly hairs or down in varying length and density, giving some species a grey felty appearance.

Stipules are lanceolate and present at the swollen nodes. Light to mid-green base colour is usual. The leaves of some species grow from a central crown to form a hummock and in others the leaf stems grow long creating a dense mat covering the ground. Autumn colour is evident in many and indeed a few maintain this red or russet hue for most of the year. The height of the foliage can be as tiny as 1 in. (2 cm) or up to 24 in. (60 cm) or more but these massive proportions are confined to one species.

Blooms

The flowers contain ten stamen but only five are fertile and usually contain both sexes. Five nectaries are present as in *Geranium*. More than two individual blooms are usually seen on each umbrel but some species do only have solitary blooms. Five petals, unequally placed, are, in size, anything from 0.125 in. (2 mm) to 1 in. (2.5 cm) or even larger and depending on the species, generally on long stems. In colour they may be white to pinks, mauves and magentas with or without blotching, spotting or striations. None possess very vibrant shades or colours thus relating to the gentle hues of the mainly fine, feathery foliage composition and suggesting a delicacy all too often overlooked in the garden.

CULTIVATION

The species and their derivitives from the drier warmer climates are perhaps better grown in pans or pots in the alpine house or cool greenhouse, at least for the winter months. Compost for these should be of a very open and free draining, loam-based type. Clay pots seem to suit better but always remember to soak the clay pot for at least four hours in water before planting up. This is to ensure the porous pot will not absorb moisture from the compost initially and cause the roots to dry out which would cause the compost to shrink from the sides of the pot, moving the roots and perhaps breaking or damaging them. Compost must be kept just damp for the first few days of any potting up procedure to settle the roots and give the required moisture to enable the roots to begin to grow. Those species and varieties that are not too fussy and planted in the open ground will benefit from a temporary overhead covering of glass or similar during excessively cold or wet winter spells as a preventive measure against crown rot.

To simulate the natural growing requirements a fairly large percentage of coarse horticultural sand or grit dug into the top spit of ground is necessary, unless the soil is known to be of adequate drainage properties, together with a small quantity of well rotted compost. A collar of shingle or gravel around the neck of the plant will keep the vulnerable area of the root crown well drained and also deter slugs and snails as well as helping to avoid damage or marking caused by heavy rain splashing soil onto the plant. The use of proprietary strawberry collars is a good idea if this type of damage is likely to be experienced. A lime soil is preferred to an acid type but the latter can be rectified by adding lime in the autumn. Care should be taken when digging so that the fine roots which are often quite near the surface are not damaged – a small fork is the best tool to use for turning the ground and incorporating additives.

For planting new beds or additions of new plants, April seems to be the best month if the ground is not too wet with spring rains. Prepare the area well and plant the specimen to the same level as it has grown in the container, taking notice that the crown is not situated below ground level. Water to settle the plant after which, normally, in a garden, the plant will not require further watering unless a drought is threatened for the period just after planting. An open, sunny position is required and many forms will enjoy the conditions offered in a rockery, scree or alpine landscape especially if a slope facing the sun can be provided. The base of a wall, facing south for preference, or a stone wall in which the plant can be planted during construction will be favoured. Some species are willing to grow happily in small nooks between the rock-face or in pieces of tufa rock. This lightweight porous rock can easily be drilled or scraped to make a hole large enough to accommodate a mixture of compost and shingle for the plant's roots and anchorage, the roots will soon discover the maze of tiny apertures in the rock. Large pieces of tufa can be planted and placed on the patio or balcony for an interesting feature.

Constructing a bed purely for *Erodium* will supply any garden with an interesting feature because the best method would be to design a raised bed with a stone or brick retaining wall, providing some slope from back to front with the entire face pointing south as an ideal. Use plenty of rubble as the base of the prepared soil which should be fertile and open as mentioned earlier and a 1.5–2 in. (4–5 cm) layer of grit or shingle over the whole surface. Only the smaller species and cultivars could be accommodated in this situation but as most are in this size category a very variable display could be created. Those needing some overhead winter protection could either be covered individually in place or if their pots were sunk in the bed these could easily be lifted to be housed in the greenhouse. Where the spaces were vacant, January to April flowering bulbs could be grown.

Plants may be purchased at a reasonable price although normally a little more expensively than *Geranium*. Never buy bare-rooted specimens. Check that the crown of the plant is undamaged and is situated above the soil level and that the top of the compost is not covered with moss or lichen, etc. – this hints that the plant has been potted for a considerable time and been kept in a dull damp place. On the other hand, don't buy plants if the compost in the pot is loose and very fresh, as this signifies that the plant has not been potted long enough to ensure a healthy rootstock. The foliage should be lively and not showing signs of rot or pests. It is better to buy a semi-mature plant, perhaps with some flowerbud formation showing than a large specimen in full bloom. The former will tolerate replanting much better and the buds will almost certainly open to give a full season's flowering period.

It is important to keep the plants, beds and pots free of plant debris by removing all spent blooms and dead or dying foliage. Staking may be needed for the lax or tall growers but twigs alone should be adequate support.

PROPAGATION

Seed

Because *Erodium* seed is still in the casing when it detaches itself from the style or rostrum it is recommended that the seed is removed from the shell before sowing unless the seed is fresh then the outer casing may be left on. Sow seed as soon as possible and follow the procedure as for *Geranium* adding 0.25 in. (0.5 cm) of dry horticultural sand on the top of the compost. If sowing with the outer casing remaining, press the pointed end of the seed unit just into the soil and allow the spiralled awn to screw itself into the compost when humidity is high. This is the quickest method if many seeds are to be sown at one time. Stratification is not often necessary but if sowing during February to May gentle heat will be beneficial for most species and varieties including the annuals.

The cotyledons will appear from about seven days depending on variety and heat. In some cases the cotyledons are reflexed, notched at the edge, heart-shaped or oval and not always symmetrical. Hairs may be evident on these mid to dark green leaves. Quite often the stems of the seedling are red or light brown above the soil. After a day or two the first two true leaves appear showing the ultimate shape of the foliage. It is at this stage when potting up of the seedlings can commence. Carefully lever the seedlings from the base with a small kitchen fork, holding them by the two cotyledons. Pot individually into pots containing a 50:50 mixture of horticultural sharp sand and potting compost such as John Innes No. 1. This must be well mixed with the hands to aerate, put into the pot without ramming down too firmly, watered well and left to drain. Keep in an area where the temperature is not less than 45°F (7°C). Do not plant out in the open until frosts have ended.

With annual types, seed propagation is the usual method. The tap-rooted kinds will prove easier by seed also, because dividing or taking cuttings may cause irreparable damage if the task is undertaken by the inexperienced.

Division

Dividing *Erodium* is not recommended; it is possible, but great care must be taken. With the fingers or a small hand fork carefully tease away a small piece of healthy rootstock with a growing stem. Only take small pieces at any one time and only from a plant large enough to recover if any damage is done. Pot on the portion, label and leave in a sheltered place until the roots have begun to grow into the new potting compost, then the new plant may be planted in its permanent place.

Cuttings

Ideally in April, break or cut with a razor-sharp knife a side shoot or scrambling shoot, with or without a heel and from mature wood. Write out a label with all the relevant details. Trim off the heel tip or trim to below a leaf node then follow the procedure as for *Geranium* but with a cutting mixture of 75 per cent horticultural sand and 25 per cent sieved or rubbed sphagnum moss peat. By placing the cuttings on bottom heat the rooting will be speeded up, otherwise rooting will take from about five weeks depending on temperature, light and season, etc. A cool or heated greenhouse or a cold frame will provided a suitable alternative during the late spring or summer months.

Pot up when rooting signs are evident into a mixture of John Innes No. 2 potting compost to which has been added about a quarter by volume of a drainage medium such as perlite, grit, vermiculite, etc. Clay pots should have been soaked in water for a few hours and should have a layer of broken crocks at the base over the drainage hole to assist in draining the compost. This is because clay pots are sited flat on the ground blocking off any drainage route so another area to which surplus water may run is essential. A bed of pebbles or other means on which to stand the clay pot is suitable. Some very old clay pots have holes at the bottom of the pot sides instead of at the base, these are perfect and allow the water to escape when needed. Crocking takes up valuable compost space so any way to eradicate the need to crock is worth considering. Crocking also encourages insects and some pests to take up residence in the rubble provided at the base of the compost, a hazard to be avoided where possible. Plastic pots often have many holes at the base and also a slight ridge enabling the pot to stand fractionally clear of the ground so that crocking is not required, but the *Erodium* does enjoy growing in clay pots. The more even compost temperature and the porous nature created by clay associate with the natural situation of open, airy composts so the little extra effort of growing them in clay pots is worthwhile.

Root Cuttings

Some species can be propagated by this method. March is generally the most successful month. Follow the basic instructions for *Geranium* but the cuttings should be no more than 2 in. (5 cm) in length. Pot up as advised and leave in a cold frame or similar situation until the next spring before planting out, after the last frost. Don't forget to label.

Layering

Layering is possible following the procedure for *Geranium*. It must be mentioned that this method of producing new plants from mature plants is not as highly successful as other methods of propagating *Erodium*.

HYBRIDISING

There are only relatively few cultivars in the genus which points to their being more difficult to cross pollinate and set seed.

However, recently one or two worthy hybrids have been introduced. By all means attempt to hybridise following the same procedure set for *Geranium* remembering that there are only five fertile stamens. Placing protective coverings over the blooms will be difficult with the more diminutive species and cultivars but even a chance crossing of merit will boost this charming member of the tribe.

PESTS, DISEASES AND DISORDERS

Again, fairly trouble free. The section for *Geranium* pests and diseases can be taken as general for *Erodium*. Thankfully, the latter very rarely suffer from forms of rust, but botrytis is more of a problem. Keeping the surrounding areas free from plant litter and removing spent blooms, etc. will minimise this complaint and also encourage the plant to flower longer, extending the flowering period which is normally June to September. Overhanging trees and shrubs, particularly deciduous varieties, should be kept trimmed back and fallen leaves removed from the bed. Crown rot can be avoided by using a collar of gravel as previously mentioned.

EXHIBITING

It is unusual to find an *Erodium* class in normal show schedules. During the last few years they have been included in *Pelargonium* and *Geranium* show schedules as well as in Alpine show schedules and have created quite favourable response from the general public and the enthusiast alike. Showing *Erodium* is not difficult particularly with the smaller specimens which adapt readily to pot culture and are often only grown in pans or pots in the greenhouse or alpine house so these can be transported to shows with little effort.

Cleanliness of plant and pot is essential and if clay pots are used this task must be very thorough. A top dressing of gravel or shingle will help to show the plant off well. Judges will look for stocky, healthy growth, good foliage and flower formation and colour and, an important point, the plant's size must be in proportion to the pot, an easy thing to forget when exhibiting specimens of a naturally small stature. Often half pots will be more suitable to plants of small habit but check show schedules first to ensure half pots are allowed.

Floral Art classes sometimes request *Erodium* blooms in flower arrangements and due to their usually petite size they are perfect for miniature or petite arrangement classes. The stems are on the whole very thin and delicate so be gentle when fixing them into flower fixing foam or other types of holders. Always condition the flowers after picking and before arranging by standing them up to the flower-heads in cold water for at least half an hour; this procedure should be adopted on all blooms within the family and the blooms should then last seven days or more depending on room temperature and vase hygiene.

ERODIUM PLANT LIST

E. alpinum: Flowers purple, six to eight in the umbel. Smooth, cut leaves. 12 in. (30 cm). Hardy. B

E. botris: Annual. Prostrate with small pink blooms deeply veined at base of petals. B

E. chrysanthum: Pale yellow blooms, silvery foliage. Not always hardy without cover from winter damp. 4–6 in. (10–15 cm). E

E. corsicum: Pink or white blooms on neat mats of green foliage. Half hardy. 3 in. (8 cm). E

E. hymenoides: White flowers with mauve-brown blotch on top petals. Leaves lobed and grey-green. Perhaps only half-hardy. 5–8 in. (13–20 cm). B

E. manescarvii: Blooms large and purplish-red. Long fern-like leaves of green, deeply cut. 12–24 in. (30–60 cm) Hardy. B

E. petraeum: White, pink or violet with two upper petals having a black patch at base. 3–6 in. (8–15 cm) Hardy. B

E. romanum: Biennial. Spring flowering. Flowers purplish with many in the umbel. Feathery leaves. 6–9 in. (15–23 cm) Hardy. E

3

Monsonia

This genus is named after Lady Anne Monson whose family lived in the South African Cape. She was a person of enormous botanical and natural historical knowledge and also was a frequent correspondent of Linnaeus.

The first *Monsonia* was introduced by Mason in 1774 and this was *Monsonia lobata*.

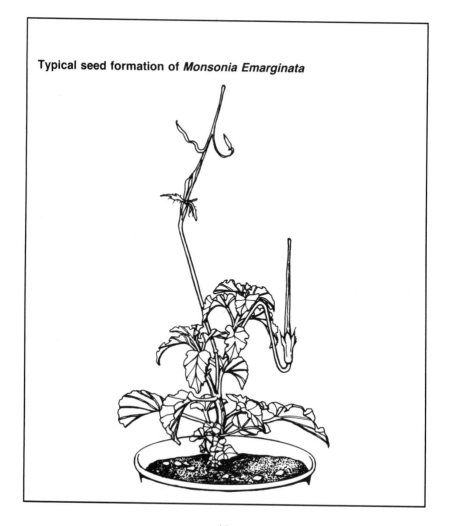

Typical seed formation of *Monsonia Emarginata*

The obvious evidence that *Monsonia* belong to the family Geraniaceae is the typically beak-shaped fruit but there the visual likeness ends. The fruit breaks on ripening and is exploded into its five separate mericarps and thrown by this force when the style or rostrum dries. Each seed is spiralled through the air, propelled by the hairy awn, and onto the soil where it awaits the suitable conditions to begin its corkscew action into the ground.

It is thought that there are about thirty-five to forty species which are arranged into seven sub-genera – these are found mainly in western and tropical parts of Africa and a few in Asia.

Available records tell of the plant's use as a remedy for diarrhoea and dysentry and it is believed that some were sent to England by the sackful during the last century for trials relating to their medicinal properties. The processed resulting tincture was evidently then transported to India for use during military action but no firm news on that score is available!

Monsonia are not often grown in cultivation and are quite difficult to acquire. To purchase a plant from a nursery may be almost impossible and the only way to obtain a new plant will be to produce it from seed. Seed is rarely on sale but occasionally is offered within seed distribution lists of specialist societies or groups.

They are bushy, branching herbs or sub-shrubs and are either annual, perennial or sometimes bi-annual, which are mainly short lived. Their habitat is almost exclusively among grass in warm dry areas, sometimes growing at high altitudes. *Monsonia* survive in the most inaccessible and arid places and are nourished by the minute particles of moisture in the air or by the rains which can be few and far between, arriving either as a splash or as a deluge according to the seasonal climate.

ROOTS, FOLIAGE AND BLOOMS

Roots are in the main strong, sometimes twisted, vertical tap roots to equip the weedy or weak-looking top growth with anchorage as well as a capability to store water and nutrients. This root is often woody or corky in texture and the root hairs are barely visible. Such tap roots, being at times very long, and occasionally branched, make it difficult to transplant *Monsonia* as mature specimens.

Both foliage and stems possess fine or strong hairs in varying density and length. Leaves are sometimes mid to dark green, glaucous or shiny and of a general pointed oval shape with deep or shallow lobes which are sometimes deeply and irregularly cut either in teeth formation or rounded crenations. The central leaf veins are strongly indented. Size is also variable, from 1–2 in. (2.5–5 cm) long and 0.25–0.75 in. (0.5–2 cm) in a mature plant depending on the species. The leaf blade is attached to either very short or lengthy petioles which are also normally covered in fine hairs. Stipules are present at the petiole and the stem axle is formed in clusters in some species.

The foliage is arranged in whorls or singly or opposite on the stem and has the power to close together in dry conditions, thus guarding against a massive loss of moisture – it would cause some stress to the plant if the moisture level was very low in the cells of the plant. Main stems may be as short as 2 in. (5 cm) or as long as 24 in. (60 cm) and can be found sometimes scrambling on the soil surface. The fairly short main stem is corky or woody with three to six erect stems growing from this in some species.

The blooms are usually very handsome and have fifteen fertile stamens. They comprise five heart-shaped or round petals, often notched and formed in a regular pattern to create the neat bloom. Colours are cream, white through to pink and lilac and have darker veining or a deeper hue at the eye in some instances. It is a pity that some of the blooms are short lived, in some cases less than a day. In Great Britain the blooms appear during May, June and July. After flowering the pedicels often bend in a spectacular manner at the nodes, as they lengthen and give the appearance of a knee and ankle joint on the stem which bears the seed capsule. The fruit or seed head contains five seeds and is dispersed with the long feathered awn attached to the inch (2.5 cm) long mericarp. The seed is varied in colour but is usually mid-brown and dull.

CULTIVATION

Bearing in mind the natural habitat of *Monsonia*, the fact that they are rarely grown in cultivated situations and that little experience has been logged on their cultivation it may be that experimentation will reveal a system totally to their liking. However, a good loam-based compost such as John Innes

No. 1 with about a quarter proportions of horticultural sharp sand or small shingle can be recommended. Clay pots seem to be preferred to plastic. If the tap root is too long to be accommodated in a conventional pot then a piece of terracotta land-drain pipe or similar can be placed in the pot, and the whole filled to the top arranging the compost around the rootstock as filling takes place. This will allow the tap root to be potted safely and encouraged to grow on naturally which is preferred to any cutting or bending of the tap root. A top layer of sharp sand or chick grit will be beneficial and guard against stem or crown rot. Watering should be minimal and the compost kept on the dry side except at the flower-bud forming stage. Offer a half strength liquid feed during the growing season, of an organic type if possible. Some species do die down during our winter months and a careful inspection should be made before throwing the plant out as dead. *Monsonia* require a warm greenhouse environment for most of the year and will not tolerate freezing temperatures. They need excellent light and enjoy basking in the sunshine when acclimatised.

PROPAGATION

Rare plants are usually difficult to propagate, this is normally what makes them rare and keeps them in that category. *Monsonia* are no exception to this belief. Seed is the most successful way of producing additions to a collection. These should be sown as soon as possible after collection. Remove the outer casing from the seed if the seed is old. The casing and awn may be left intact on fresh seed – this way the danger of damage is reduced and the seed allowed to screw itself into the compost. Use a 50:50 mix of good seed compost and horticultural sand which should be well soaked and allowed to drain before seed planting. Label the batch. Cover the seeds with a sprinkling of fine sharp sand and place on bottom heat at a temperature of 65–70°F (18–21°C) unless a summer sowing is made.

The seedlings will emerge in about three weeks, sometimes longer, and will have cotyledons of a oval shape with a pronounced central rib. They are usually green in colour and the stems at this stage might be reddish-brown or green depending on the species and on heat and light availability. When two true leaves develop pot up singularly into a small crocked and soaked clay pot using John Innes No. 2 with equal parts of sharp sand mixed throughout. Give water gently and sparingly. If roots of seedlings become entangled it is far better to pot them up together than to pull them apart at this stage, waiting until the seedlings have grown away well before trying to separate them and pot singly. Shelter them from sunlight for a day or so, gradually bringing them into full light as they become acclimatised. Finally house the plantlets in a well–lit warm greenhouse or conservatory. They may be placed outside on warm summer days but avoid strong winds which will damage the delicate blooms.

Division

This type of multiplication is not recommended for *Monsonia*.

Cuttings

Stem cuttings may be tried using the recognised method of cutting just below a node and placing in a watered and well drained mixture of 40 per cent sieved sphagnum moss peat and 60 per cent perlite, vermiculite or horticultural sand with a shallow layer of fine sharp sand on the surface. Label, then place uncovered on bottom heat at a soil temperature of 65°F (18°C). Rooting may take a few months and might not result in a good percentage of take. Root cuttings may also be tried and depending on the species may be more successful. Follow the procedure as for *Geranium* but keep the roots in a warm greenhouse. It is not practicable to try the tap rooted ones for this method due to the damage to the parent plant and the fact that the results might not furnish enough plantlets to warrant destroying a good mature plant. Seed is therefore the most foolproof way of producing new specimens.

HYBRIDISING

To date there are no man–made varieties (cultivars) of *Monsonia* but individuals are continually trying. To ensure one's own species plant sets seed it is a simple task to observe the plant in question for the production of pollen and receptiveness of the stigma. Collect the pollen onto a small brush or similar such as a cotton bud and transfer the pollen onto the pistil of the same or another plant. Seed does not set too readily in a number of species.

PESTS, DISEASES AND DISORDERS

There are relatively few, the main problem being rot, both at the seedling stage and in the mature plant due in the majority of instances to over-watering. A fungicide may be added to water but the easiest prevention is careful, limited watering at all times especially in the colder periods of the year. Greenfly can be controlled with purchased preparations or by introducing predators such as ladybirds to the greenhouse. Occasionally leaves may be edged with brown markings often killing that portion of the foliage; this is sometimes due to draughts but also due to over-feeding, strong light or strong sunshine when the plant is still young.

EXHIBITING

There are never any individual classes for *Monsonia*. The only place in a schedule they may be shown in is the rarely provided class 'any other Geraniaceae'. Judges, generally, would have little knowledge of the plant and so overall points such as cleanliness of pot and plant, good formation of plant, blooms, leaves, good condition, as many blooms and following buds as possible would be the only guidelines. A top dressing of sharp sand or fine shingle would give the exhibit a 'cared for' appearance.

MONSONIA PLANT LIST

M. augustifolia: Usually an annual but has survived longer in greenhouse conditions. Long growth sometimes as a prostrate or upright habit. Leaves, long, oval with serations, green with some fine hair covering which are also present on stems and peduncles and pedicels. Flowers two or three to the umbel, whitish with a blue tinge, occasionally, cream or white. 20 in. high at maximum (50 cm). Best perhaps grown from seed. E or B with care.
M. emarginata: Grown as an annual but may live longer. Foliage, wedge-shaped, dull green having shallow crenations, sometimes wavy at margins. All stems slender and hairy. Peduncles single flowered with blooms of pale yellow, edges waved. Southern Africa. 8–10 in. (20–25 cm) From seed B.

4

Sarcocaulon

Plants of the *Sarcocaulon* genus are all succulent, often spiny and have a wax-like coating or bark on their usually thick main stems and branches. The word *sarcocaulon* is derived from the Greek *sarco* – fleshy – and *caulon* – a stem. In its native Southern Africa it has gained many nicknames, the most popular being, 'bushman's candles', 'poison thorn', 'hell thorn', 'stomach thorn', and 'pork crackling bush'. All these describe the nature of these spiny, shrubby, low growing perennial plants. Because the translucent wax of the stem bark will burn readily and incidently give off thick smoke, often the bark from dead plants, discarded by the action of the sun and wind, will eventually form into round orangey-brown pearls of wax. Being translucent helps the plant to survive the long periods of searing sun and the periods of drought.

Many of the fifteen or so species are single stemmed and much branched, varying from 3 in. (8 cm) to 60 in. (1.5 m) in height and with the scrambling species perhaps up to 24 in. (60 cm) spread. They are grey-green or occasionally lightish to very dark brown in colour. The spines also vary in length, from just a short stubby projection to a strong spine of up to 2 in. (5 cm) or more in length.

It was during the eighteenth century that they were first introduced and also illustrated, but grouped firstly under *Geranium* and later under *Monsonia*. About a hundred years later the grouping was revised and the latest work on classifying the genus was undertaken in 1979 by Moffett who split the genus into four sections, the listing in these sections being determined by the shape of the leaf edges.

They are found in harsh desert lands and sometimes in the sandy diamond areas, often in quite windy places of plain-like terrain. Some inhabit coastal runs and a few have adapted to the rocky and pebbled ground-cover where the sea mists and dews are often the only means of moisture to the plant. Conditions are, at times, severe and so cold that unbelievably *Sarcocaulon* have to tolerate some freezing temperatures at night. This they have acclimatised to but they will only survive if the plant is in a compatible condition and state at this time; even so, many succumb to these extreme elements. This compatible condition is almost impossible to simulate in cultivation and would be impossible to detect anyway, so *Sarcocaulon* must, in the Northern hemisphere, be classed as tender from our late autumn, through to winter and spring and be safely protected in heated greenhouses until all risk of frost has past.

In the main, their habitats are away from other plants and in the open without overhanging trees or rocks, etc. so that intense light is at hand at all times. These points are rarely descriptive of a greenhouse, but it is worth considering providing extra space to house *Sarcocaulon* to ensure fit and healthy plants. In their natural environment the growing period is normally only about three or four months of the year; the remaining time the plants are totally dormant having lost all green parts before the resting period. Sometimes the dormancy periods last for more than one year. It is amazing that the plant will begin and continue to grow foliage and new roots, bud, then proceed to bloom, set seed and distribute seed in only a quarter of a year. Perhaps this is one reason only a handful of people include this interesting plant of the Geraniaceae in their collection – succulent collectors are the main enthusiasts it seems. However, the reward of raising and caring through one year alone is three or four months of absolute delight and ever-changing interest. It is an advantage to own plants which need a dormancy period for they may be tucked underneath the bench and almost forgotten making space for seasonal plants.

The plants may be purchased from cacti and succulent nurseries at odd times but it is a case of keeping a hawk-like eye on lists and catalogues or asking to be included on waiting lists for any to become available. A fairly mature plant will be expensive compared to the other members of the family. Seed may be included in the specialist societies' seed distribution and this will be the cheapest means of obtaining new plants.

ROOTS, FOLIAGE
AND BLOOMS

In the majority of species the roots are shallow, fairly sparse and woody. Any tap root will be short and often horizontal – this is to assist the take up of moisture from surface fog and mist. If swollen roots are evident this gives clues to the area where the plant would reside in the wild due to the conditions of intermittent but heavy rainfall necessitating a water storage function in the root structure. *Sarcocaulon* do not transplant well and because the root system is not large, anchorage could be a problem and a cane or two used to prop up the newly potted plant may be required as a temporary measure. If plants are purchased with the roots dry it is important that on receipt of the plant a gentle spraying of warm water allows the roots to begin to re-function before the specimen is potted up.

The foliage is often almost non-existent, sparse and small. There are two types of leaf, either long or short stalked. The long leaf stalk or petiole remains on the stem after leaf-fall as a spine. The leaf shape is sometimes oval, sometimes wedge-shaped at the base, either segmented, lobed and dissected and cut or it may be heart-shaped or kidney-shaped depending on the species. The leaf colour is mainly grey-green but some possess light green foliage – in size, from 0.25–1 in. (0.5–2.5 cm) long and half as wide and occasionally clothed in fine hairs either below or above the leaf blade. The leaves are able to close laterally at the centre rib, which is often prominent, to protect the leaf face from the hot sun. It is not unusual for there to be an absence of foliage at flowering time.

Stipules are spiny or may be found in clusters of small spiky tufts.

The typical succulent form and light grey-green or light to very dark brown stem can be very smooth to the touch or quite rugged or woody in a mature plant.

Flowering periods can vary but are usually spring and summer. Each bloom has five petals evenly spaced and having fifteen anther-bearing stamens, all being fertile. White to soft or sulphur yellow, rose to salmon pink and magenta often with a large paler or white eye are the range of hues in the flowers which are about 1.5 in. (4 cm) across, having an appearance of crumpled damp coloured tissue – in some the petals are notched at the tip. The short life of individual flowers is balanced by the profusion of blooms on a well-grown plant. It is difficult to state a flowering period on specimens grown away from Southern Africa because the length and spirit of the dormancy period will dictate this but the blooms will follow the foliage production in most cases. When a plant has set seed the ripe fruit will contain five mericarps of one seed each. The awn is gently coiled and plumed to assist in the springing of the seed away from the parent plant.

'Brunswick'

Pelargonium inquinans

Geranium clarkei 'Kashmir White'

'Henry Cox'

Deacon 'Lilac Mist'

Rollinson's Unique

'Mr Wren'

Unique hybrid 'Carefree'

CULTIVATION

As with *Monsonia* it is apparent that only others' experiences will assist in the ideal cultivation of *Sarcocaulon* due to the lack of available cultural advice to date. Ample light is vital especially at the time the plant emerges from its dormancy period. A temperature of not less than 40°F (5°C) in a glasshouse is required and temperatures in excess of this for the growing period. Low humidity should be aimed for, apart from the times when a very fine spray at the start of the growth cycle with tepid soft water will be welcomed. Water at all times sparingly, increasing slightly at flowering and fruiting times when a quarter strength liquid feed may be given.

Compost must be of an open nature. A good quality loam based potting compost of John Innes No. 1 specification with added grit, chippings or perlite to a ratio of 4 compost 1 drainage material. The size of the grit or chippings should ideally vary with each species depending on the natural terrain in the plant's habitat. A top dressing of the appropriate chippings will be beneficial and clay pots do seem to be preferred to plastic. Half pots or deep pans are in some cases quite adequate. The support of twigs or short canes may be required and removed when the plant has formed new roots into the surrounding compost.

Always place *Sarcocaulon* plants with plenty of room between them. Never allow them to stay in an overcrowded situation for long. During dormancy spells the plants can be placed in a less important part of the greenhouse. Do not water during the depths of dormancy but do check the plants at regular intervals for problems and also observe when the specimen is likely to break dormancy.

PROPAGATION

The sowing of seed, which should be as fresh as possible, is the surest way to propagate *Sarcocaulon* successfully. Remove the mericarp if the seed is old and still in its outer casing. Germination is quite speedy with fresh seed, between twenty-four hours and fourteen days after sowing. Older seed will vary both in germination time and in rate of germination. The compost for seed raising must contain adequate drainage material of a smaller size than for potting mixtures so chick grit or seed grade perlite is ideal mixed with good quality loam-based seed compost to equal proportions. Place the well-mixed medium into, preferably, a soaked clay pan or shallow pot, sink in water until the surface of the compost is just showing damp, remove and leave to drain. Press the surface lightly and place seeds on the compost. Cover with a sieving of dry, fine horticultural sand or fine chick grit. Label and place on bottom heat, apart from the summer months when the pan can be placed under the top benching of the greenhouse. The cotyledons sometimes appear red and are oval in shape, folding slightly along the centre particularly in full sunlight. When the leaves are large enough to handle prick out the seedlings into small individual pots using a loam-based compost of John Innes No. 2 or equivalent with equal proportions of grit and shingle. Top dress with the same. Water round the edge of the pot and leave in the shade for a day or two to recover.

Stem cuttings may be attempted either from pieces cut straight across at any point of the stem. Use Flowers of Sulphur to dust the wound on the stock plant. Insert the cutting into a compost mixture as with seeds, including the top dressing of fine sand, and water very sparingly. Further watering should not be required until the new roots have begun to form. When this has occurred pot up as with seedlings. Success might be limited but if seed is unobtainable cuttings may be the only way to produce new plants.

HYBRIDISING

Seed does not always set readily in greenhouse conditions and a little human transferring of pollen is usually needed. At present no cultivars of the *Sarcocaulon* genus have been created.

PESTS, DISEASES AND DISORDERS

Mealy bug and scale insects may attack the plants around the stems. Pick off individually even though the task will be laborious. Using specified sprays may cause damage to all parts of the plant. Sometimes root mealy bug may be residing in the pot of compost. Washing the roots will remove the pest but the problems of potting up again will remain. There are preparations to combat this pest and other soil pests but it is important to use these with care and only use on one plant initially if doubtful of the effect on the plant. It is fairly safe to use preparations specially formulated for cacti and succulents but do take care both with the plant and any humans or animals. Remember, prevention is far better than having to use chemical preparations. Wherever possible give organic methods a fair chance. Occasionally greenfly will attack young leaf growth. Remove with the fingers when the infestation is small or carefully spray with a preparation specifically for use on tender ornamentals.

Blackening of any portions with softness of the tissue will denote decay and to prevent spread it is advisable to cut away the affected part back to a wholesome area. Use a clean razor-sharp knife and seal the wound with a dusting of Flowers of Sulphur or powdered charcoal. Both cool conditions and over-watering can cause this problem.

EXHIBITING

Where specialist shows provide a class for *Sarcocaulon* or perhaps for 'any other Geraniaceae' it is an unusual invitation to enter. To exhibit is the way open for growers of *Sarcocaulon*, and the other not so often grown members of the family, to introduce the general public to these plants and hopefully encourage an interest. Often cacti and succulent shows have classes for 'any other succulent' and *Sarcocaulon* plants may be entered in these classes also.

Tidiness and cleanliness of both pot and plant, good shape, condition and colour of foliage, blooms and buds will be what the judges have as a guideline. Top dress with an appropriate medium and add a name label. Make sure you have read the schedule and show rules and adhere to them.

SARCOCAULON PLANT LIST

S. crassicaule: Thick stems and branches of greyish-yellow colour forming a spiny shrublet. Leaves lobed and slightly downy. Blooms of five petals, sulphur to pale yellow. From Western Africa, growing in mountainous, rocky places. 18 in. in height and spread (45 cm). E
S. inerme: Dwarf, prostrate shrublet with short spines hardly noticeable. Leaves oval with wedge shape at base, crinkled at

edges, soft and downy. Blooms, mainly rose-purple. Found in the foothills and mountain edges of South-West Africa. 12 in. (30 cm). E

S. patersonii (syn. *S. rigidum*): 'Bushman's Candle', 'Hell-thorn'. Prostrate fleshy, waxed coated, golden to grey stems. Leaves greyish with entire margins. Flowers, rose-pink to magenta, quite large. Height 18 in. (45 cm). E

S. salmoniflora: Small, thin branches of brown bark covered stems. Leaves, folded, oval with pointed tips, fairly shiny. Flowers, salmon to pale orange. Widespread from Southern Karoo, across Orange Free State and Northern Namibia in rocky areas. 15 in. (40 cm). E

S. vandieretiae: Branches spiny, olive-green to grey. Foliage, grey-green folded heart-shape with notch at tip. Spines in rows and shortish. Flowers pale pink or white with pink shading. From Eastern Cape in valleys of sandy soil. Wider rather than taller in habit – 6 in. in height (15 cm) spread about 12 in. (30 cm). E

5

The
Pelargonium

INTRODUCTION

Within the *Pelargonium* side of the family there are four basic divisions with further sub-divisions:

1.   The Species, including primary hybrids, the Scented-leaveds, and the Uniques.
2.   The Regals, including Dwarf Regals and Angels.
3.   The Zonals. The largest number of types belong to this section and are the plants commonly referred to as geraniums, including those with either single, semi-double or double blooms; miniature and dwarf growers with either single, semi-double or double blooms; fancy flowered with single, semi-double or double blooms; ornamental foliage varieties with single, semi-double or double blooms (ornamental foliage types are present in plants of all divisions except true species); and lastly the modern F1 hybrids.
4.   The Ivy-leaved, including hybrid Ivy-leaved, with either single, semi-double or double blooms.

As we have already seen, the name *Pelargonium* comes from the Greek for stork, and although there is little difference in shape between the beak of a crane (the *Geranium* is also known as cranesbill) and that of a stork, the name has remained with this genus over the last few hundred years.

The *Pelargonium* is, without doubt, horticulturally the most popular and important genus of the Geraniaceae family. The main botanical difference between *Pelargonium* and *Geranium* is the presence of a nectar tube, or spur, which is located at the base of the upper petals (sepals). This continues along the flower stalk (pedicel), often kinking the stalk to give the appearance of a limb joint, and creating a visible bulge. A more detailed observation may be obtained if a large flower is sliced along the pedicel, through the petals. In most cases, this

Pelargonium triste.

will reveal a long nectar tube, although in some species the tube may be quite shallow and the kink before the main pedicel not so apparent.

When a flower is fertilised the ovary swells and fruit forms in a set, usually of five mericarp, each mounted on the rostrum at the base, curling upwards to the tip of the 'beak'. Each remains attached by a coiled and feathered awn for perhaps a time, until the weather conditions are right for the distribution of the seed. The long hairs of the awn look like feathers or plumes when unfurled and act both as a spring mechanism to launch the seed into the air, and as a parachute to float the seed often quite a distance from the plant. The rostrum remains on the plant until it dies and falls away.

The elliptically shaped seeds are smooth or marked in a net–like mode, but have minute bristles which act as a kind of anchor, holding the seed on the ground. The coiled feather corkscrews itself into the soil with coil and uncoil movements which depend on the heat and moisture at soil level. The seeds are generally mid–brown to brownish grey in colour, but sometimes rust coloured seeds may be found. They range in size from that of a pin–head to 0.25in (5mm).

The first *Pelargonium* is thought to have been introduced to Holland in the seventeenth century; this was perhaps a plant of *Pelargonium zonale*. One of the first *Pelargonium* to reach the British shores was brought from the Cape of Good Hope via Europe and into the possession of John Tradescant in 1631, first passing through the hands of M. Morin of Paris. The following year it was blooming in the garden of the Tradescants in Lambeth, London. The plant was then called *Geranium Indicum nocto odorato* ('Sweet Indian Storksbill' or 'Painted Storksbill'). At that time all *Pelargonium* were classed under *Geranium* and it seems right to assume that the *Indicum* tag derived from the fact that ships of plant-hunting expeditions returned from the east by way of the Cape, and stopped off for stores or more specimens. Once back in Britain, the whole cargo was thought to have come from India. The last two words, *nocto odorato*, mean 'night scented' – this, at least, is true. This plant is now known as *Pelargonium triste*.

It was not until around 1720 that Jacob Dillenius, it is generally regarded, first suggested the term *Pelargonium* for some types of the plant then known as *Geranium*. The French botanist Charles Louis L'Heriter used the name *Pelargonium* in his unfinished manuscript describing eighty-nine species of the *Pelargonium* in 1789. The Royal Botanic Gardens at Kew,

England, are reported to have had a plant of *Pelargonium cucculatum* in the late seventeenth century. The arrival of more speciments from afar saw the beginnings of hybridisation. Since that time this diverse array of easy to grow, reliable and generously colourful plants has continued to give much satisfaction and pleasure. Yet the species has, in some respects, eluded the hybridists who have not to date developed a true yellow, a black or a scented flower, although these are all present in the natural species.

Most of the 250 or so *Pelargonium* species are found growing wild in Southern Africa. Smaller numbers are also found in Madagascar and East Africa, through Arabia, Syria and Western India. Australia and Tristan da Cunha can also boast their presence; so too may St Helena who has, mysteriously, a solitary species *Pelargonium cotyledonis*. There have been cases where *Pelargonium* have escaped from collections and gardens in suitable climates and places of ideal habitation to develop into colonies away from their natural environment.

Some *Pelargonium* are reputed to have been used as medicines for the relief of diarrhoea, dysentery, as an astringent, to soothe sore throats and to condition the hair. There is an anthocyanin pigment called, in fact, *pelargonin*, present in some of the petals. The native women would use the scarlet stain from the petals of the red–flowered forms to paint their faces and dye fabrics. How many little girls today have been caught plucking a few petals to rub on their skin as lip paint and blusher? The blue stain from other wild *Pelargonium* is also used as a dye and paint.

The many virtues of the *Pelargonium* cultivars have made the plant extremely popular both in commercial horticulture and with the amateur gardener. So popular in fact that many national societies have been formed around the world catering for the enthusiast of the Geraniaceae family and in particular the *Pelargonium* genus. (A list of contact addresses may be found at the end of the book.) Apart from the parent societies, many hundreds of local clubs have formed in Great Britain where both beginner and expert can exchange ideas, show off specimens, and generally enthuse over this wonderful range of plants which give so very much in return for so little effort.

THE SPECIES

In 1860 an Irish botanist named Harvey divided the *Pelargonium* genus into fifteen sections as follows: *Hoarea, Seymouria, Polyactium, Otidia, Ligularia, Jenkinsonia, Myrrhidium, Peristera, Campylia, Dibrachya, Eumorpha, Glaucophyllum, Ciconium, Cortusina, Pelargonium*. Each plant mentioned later in the plant list will be attributed with its section. Many attempts have been made since to revise the taxonomy and nomenclature of the genus, the most recent being one by Reinhard Knuth (1874 – 1957), a German botanist from Berlin who published his revision in 1912. Knuth divided the *Pelargonium* genus into the same sections as Harvey but included more species in each section. More recently *Pelargonium cotyledonis* was removed from the *Otidia* section and placed in a section on its own named *Isopetalum*, making the sixteen sections known today. A group of botanists are at present working on an all-inclusive taxonomy study at Stellenbosch University in South Africa under the direction of Dr van der Walt. The study began in the 1970s and will take many years. Three excellent volumes, beautifully illustrated, of their work have been published so far. Dr van der Walt and his team are indeed fortunate to be working on the 'door-step' as it were of the 'Pelargonium Garden of the World'.

ROOTS, FOLIAGE AND BLOOMS

Roots

The three basic root forms are: fibrous, tap and tuberous.

In the main the fibrous roots are woody and springy, usually (not always) brittle, occasionally growing from a main root-like horizontal stem or rhizome from which new upper plant growth arises as a sucker-like plantlet. Generally the fibrous roots are shortish and confine their growing to the area nearby.

Tap-rooted specimens are to be found in places where water content and nourishment in the soil fluctuate and so must be stored during short starvation periods, some tap roots searching many feet below for moisture. There are not many tap-rooted kinds of *Pelargonium*.

Tuberous-rooted or geophytes have storage organs, often

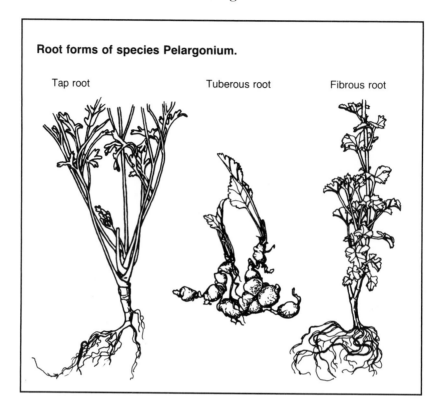

Root forms of species Pelargonium.

Tap root   Tuberous root   Fibrous root

covered with hard skins like bark and they can be quite tiny or as large as a tennis ball. They have the ability to store food and moisture for long periods of drought absorbing these during the infrequent spells of rain. If conditions are not perfect the tubers will remain in a dormant state, often for years, and begin activity when the climate is more favourable.

Foliage

Leaves can be of differing sizes, from less than 0.25in. (5 mm) to more than 14 in. (35 cm) across and usually longer in proportion. They are situated both opposite and alternate on the main stem. In texture the difference is equally compound with long or short, soft or coarse hairs covering perhaps both upper and lower leaf parts, just one leaf plain, or no hairs present at all. The latter have a waxy film over the leaf surface – these types of leaf covering are to limit water loss, to deflect the sun's rays and to prevent drying winds from scorching the foliage. To the touch some leaves are soft and flannelly, others coarse and stiff. The shapes also follow the extremes from one

Pelargonium species leaf shapes.

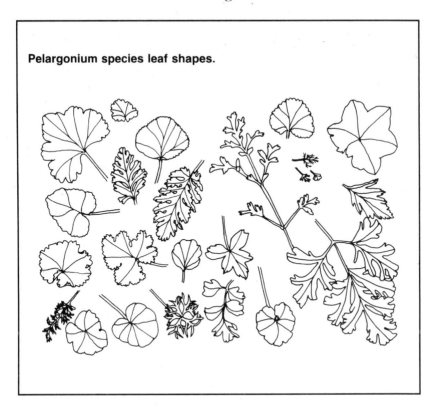

species to another or can even differ on the same plant. The simple form is entire in shape without lobes or cuts and the complicated, divided, cut or notched with leaf edges wavy or ruffled. To illustrate, imagine the shape of a normal water-lily leaf and imagine the foliage of a carrot and there you have a mental picture of these basic extremes of shape. Colour varies from dark green or light green with dark, light or non-existent zones or markings, to glaucous, grey-green or silver-grey. Occasionally, in some species the foliage takes on an attractive russet tint with maturity. Many have leaves which are highly or lightly aromatic, this scent or aroma is released whenever the plant is touched or bruised and is thought to be a natural defence against insects or animals devouring the plants. The Scented-leaved species and cultivars are at this time becoming very popular and a complete section is devoted to these interesting and useful types later in the book.

The stipules often take the leaf colouring as do the main stems. Stipules may be large in proportion to the leaf size or not so. Sometimes the stipules dry and fall away quickly and may not be apparent – their shape is either oval, triangular, pointed or round and perhaps hairy. Stems can be succulent, woolly, smooth, hairy, extremely long and scrambling or almost non-existent. The latter applies mainly to some of the geophytes where often the remains of the leaf-stalks (petioles) stay attached to the plant for many months, even years. The woody or succulent stems can possess a bark-like covering which in time is shed.

Plant height is from a few inches to a 7 ft (2 m) climber or scrambler, happy to grow through shrubs or over rocks. The plant's girth can exceed its height.

Blooms

Flowers are zygomorphic normally of five petals, occasionally fewer. Ten stamens are present, seven of these are fertile and like other members of the family have both male and female parts on each flower. The blooms are held on long or short, sometimes divided scapes mounted on peduncles again either long or short. More than one and up to seventy flowers make up the umbel-like inflorescence and one or more umbels of blooms may be found on each stem. The blooms are held on long or short pedicels which are sometimes borne on branched peduncles which in turn are situated on long or short flowering stems (scapes). Such flowerbearing structures are referred to as the inflorescence.

The diversity of colour is great. White and near black, grey, mauve, purple, pink to red, salmon, cream to yellowish-green and yellow form the basic petal colours. Flower size varies from 0.125 in. (2 mm) to over 2.5 in. (6 cm), the upper petals nearly always larger than the lower. Where fewer than five petals are present naturally, it is the lower area of the bloom that is deficient. Some *Pelargonium* species and primary hybrids have perfumed flowers, these mostly occur on plants that are pollinated by night-flying moths and so the perfume is only given off at nightfall. To attract other pollinating insects some flowers possess exquisite veining and blotching often of very bright contrasting colours. Nature has adapted the size and shapes of blooms and nectar tube in each species to accommodate the many types of insect, etc. that share the same habitat as the particular species of *Pelargonium*.

The Inflorescence of the Pelargonium.

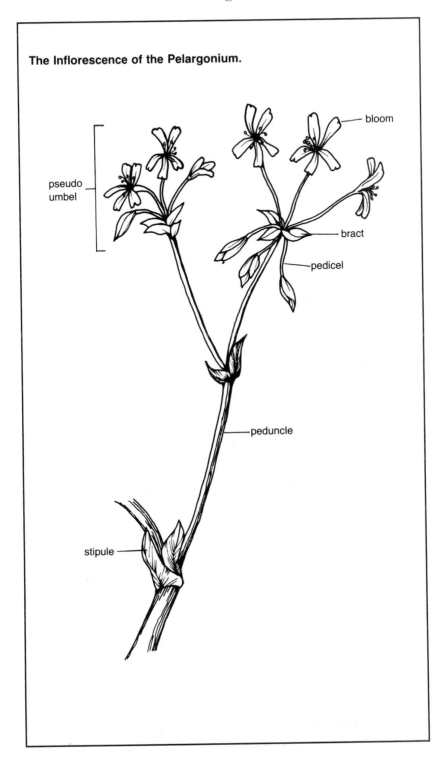

pseudo umbel

bloom

bract

pedicel

peduncle

stipule

CULTIVATION

It is important that the grower acquaints himself or herself with each species' natural habitat to enable a good pot-grown specimen to be produced, because each species is so different from another regarding likes and dislikes. Having said that, don't assume that on the whole they are difficult to grow. Ninety per cent are simple but it is the remaining ten per cent which create the challenge and these will be dealt with individually in the plant list at the end of this section on Species.

As mentioned almost all come from the Southern hemisphere, from places which are generally of free drainage, little humus from other plant life, minimal shade from the hot sun and where rains are usually few and far between. Compost should, therefore, be of a loam-based type with added horticultural coarse sand or grit, perlite or vermiculite to a percentage of one quarter drainage material to three-quarters of good quality potting compost such as John Innes No. 2. As with others in the family a clay pot, cleaned, soaked for a few hours and crocked in the base will be preferred. It is admitted that plastic is easier to clean, lighter to handle, more durable to use and store and so, as most who have grown *Pelargonium* grow quite a number of specimens, the use of plastic pots does seem more sensible. Most *Pelargonium* types will take quite happily to plastic pots, it is a case of definitely not giving too much water due to the non-porous pot.

A word here about the use of cheap or 'free' containers such as paper-thin pots normally manufactured to last for one season or disposable drinking cups. More often than not these cups and pots are of a light colour or white and if held to the light it will be observed that the light will show through – this is not a natural situation for a root system and the excessive light shining through also encourages the formation of algae on the inside of the pot or cup. Sunlight will cause this type of container to become very brittle quickly, often with disastrous results when the potted plant is being transported. When the plant is heavy with foliage the fact that the bases are normally narrower than conventional pots makes keeping them standing upright difficult. Holes always have to be pierced or melted into the base – a dangerous task! The cups do not usually have a raised base so that drainage is impeded perhaps causing the compost to become water-logged if it is placed directly on a flat surface. Well-made plant pots can be bought easily and by shopping around or joining friends for a bulk purchase are only a few pence each and certainly make a plant collection much more attractive, easier to care for and maintain. Round pots are normally used but the author has experimented using small square pots for the potting up of newly rooted cuttings and discovered that, as the young roots fill the pot, those roots in round pots will follow the pattern of the pot and grow round and round, whilst those in square pots tend to grow to the four corners. With subsequent potting on the plants with the four groups of roots seem to grow more evenly into the new compost, therefore 'getting away' more strongly.

Before potting up a newly rooted cutting or young plant, mix the compost and drainage material well, lifting and sifting it through the fingers to lighten and aerate as well as rubbing out lumps. Place some of the mixture into the base of the pot, and resting the roots on top of this, carefully fill in around the roots until the compost is almost level to the rim of the pot – a very gentle press down together with a tap on the bench is enough to settle the compost round the roots. Never firm or compact the soil with a pressing down action of the thumbs, this will certainly sheer off the upper roots and could cause damage to the stem. To water, the best method is to place the pot into a tray of tepid water, allowing it to penetrate to the surface then remove the pot and drain well before placing in a shaded place for a few days until the plant has recovered. In times of excessive sun a layer of newspaper over the plant for a few days will allay any stress that could be experienced. If it not convenient to use this method then run a trickle of water round the edge of the pot giving the plant a thorough wetting and leave to drain before placing in the shade.

The question of potting on is often taken for granted or even ignored. Remember that a well-grown plant must keep growing at about the same pace to ensure a good-shaped plant. Most of the nutrients in the compost will be used up in about six to seven weeks so to avoid a check in the top growth feeding with a balanced liquid fertiliser is necessary. For the Species only a quarter strength feed will be needed by most except the vigorous types and those growing in humus-rich areas. This is best given at every watering. The brand or type of fertiliser used will be the individual's choice but bear in mind that a fertiliser high in potash will encourage flowering and one high in nitrogen will encourage good foliage. This is only, of course, a guide and it is essential that the fertiliser is balanced and contains the other crucial benefits including trace elements in the correct proportions. Occasionally vary the type or make of feed as a change for the plant. It is advisable that beginners should purchase a fertiliser of renown and not experiment regarding feeds, especially for the Species who are not, as a general rule, hungry plants but can be choosy about feeds. The careful use of added nutrients to the compost will delay the need to re-pot and as Species do not enjoy being disturbed it is better not to pot on until really needed. If a plant becomes top-heavy with leaf growth or needs very frequent watering due to the pot being full of root it is obvious that re-potting is the only solution. Never over pot a *Pelargonium*. A simple way to cause less upheaval at this time is to sink a pot of the same size as the plant has been growing in into the centre of the new pot, fill the area at the base and sides with the new mix, press down lightly, take out the smaller pot, tap out the plant (which should have been previously watered) and set it in the shaped hole; give a gentle tap then leave to soak and drain as before. Naturally if more than a few plants are ready for potting this procedure will take a great deal of time but it is often worthwhile and certainly will be if the plant just has to be re-potted whilst in bloom, a time when any potting should be avoided if possible. When removing a plant from a plastic pot, if the pot is rolled in the palms of both hands this will loosen the compost from the pot side; then by carefully inverting the plant with the main stem between the fingers the pot can be lifted off with the free hand. This is safer than inverting the pot and banging on the top edge of the bench but sometimes this has to be done if the plant is very rootbound or in a hard-sided pot.

Many *Pelargonium* species will never fill the pot with roots and so when it is apparent that the plant is lacking desperately in nourishment it is normal to carefully scrape away some of the top compost and replace with new. The easy to grow fibrous ones can be removed from the pot and a wedge of about one-eighth of the compost sliced off, just like cutting a piece of cake, then the cavity filled with new compost. This is quite a safe procedure particularly if the pot would be impractical any larger and can be undertaken each year, cutting away a different section of the root each time. Tuberous-rooted species should be potted preferably when the plant is dormant.

Water sparingly is the guideline, especially at times of low light readings and low temperatures. Some species require much less water than others – again, noting the natural growing areas and climates will be a great help. The geophytes and deciduous types must not be watered whilst at rest and the geophytes at other times, watered very carefully. When a plant is coming into bud more water should be offered, increasing as the plant comes into full flower. Many species do grow on the edges of wooded mountains or at the edges of streams and, therefore, require more water than most, but good drainage is important. If by accident a plant becomes completely dried out the best way to ensure it receives a thorough soaking is to stand the pot in a water-bath until the top compost shows signs of water, then remove and allow to drain. If the plant is stressed it may be better to place in a shaded part of the greenhouse for a day or two. Any attempt to water from the top will be useless because the water will just run straight through to the drainage holes and away, before the compost has had time to soak up the moisture. When a plant is distressed by lack of water or a system's shock it will create a state known as excess transpiration. Transpiration is a normal plant process of expelling moisture through the foliage but when the plant is upset it will perform this to the point of causing wilt which is, of course, weakening to the plant's tissue structure and could cause the plant to die or at least lose leaves or become permanently weak.

In the wild some species grow in places where overhanging rocks give shade so although most species do grow in the open those needing shade should be accommodated in the duller part of the greenhouse or a semi-shaded part of the garden in the summer. There are some whose roots are in the shade of shrubs and other plants or rocks whilst the top growth scrambles up to the light. Placing rocks or pieces of clean solid bark or driftwood on top of the compost will create similar conditions. A top dressing of shingle or coarse sand is necessary for those not liking their root-crowns damp or for the trailing stems which will rot if kept permanently damp. Even a small clump of grass provided in the pot for those who enjoy living in long grasses will be favoured but keep the grass trimmed and do not allow it to go to seed or else you may find it taking over in every pot!

Generally only the largest and most vigorous species will require pruning. This should be done before the flower buds begin to form if possible. Using a clean razor sharp knife cut the stem across with a slanted cut at a point just above a node, preferably with a growing bud evident and pointing outwards.

It is just a case of keeping the plant in good shape and also in bounds! Some supporting may be required by a few, twigs being the natural apparatus to use, tying in with soft garden string or raffia, if necessary – try to make any staking as unobtrusive as possible. If the trailing or straggly ones become too untidy, insert a semicircle of stiff wire or basketry cane into the compost (avoiding the rootstock or any plant part that could be damaged) and then tie the stems round the structure. Wires attached to the greenhouse will give support to the vine-like specimens – remember the plants will be difficult to move after a while but if the greenhouse is large or high enough a permanent trellis system can be installed and plants encouraged to grow into the roof-space as long as the plant does not take over and restrict light to other plants or drop plant debris onto plants at bench level. Such scrambling plants can be housed in hanging pots or baskets and this method will increase the capacity of the greenhouse considerably – again be aware of the problems of falling plant debris and water dripping from these containers if they are placed above other plants.

Although *Pelargonium* species will enjoy standing outside during the British summer do not forget that they are not frost-hardy (apart from one species) and must be returned to the heated greenhouse, heated conservatory or living-room for the winter and early spring months. Many growers prefer to keep their plants in a greenhouse all the year round – this is quite acceptable but it is labour saving with watering, shading, etc. if the plants are outdoors for the summer. Pots may be sunk into the ground or the plants taken out and planted in the soil, placed on a bed of gravel or similar or perhaps grouped together as a feature on a patio. If the plants are left in their pots it will be much simpler in the autumn to take them all back into the greenhouse before the frosts.

Today a greenhouse is a normal part of the amateur gardener's hobby and many shapes and sizes are available to suit most pockets – indeed, competition between greenhouse manufacturers is so fierce that most produce notable brochures and have advice centres to assist in the purchasing, placing, erecting and after-care, including many gadgets and appliances. Books have been written on the subject and these together with manufacturers' brochures may be studied and, therefore, it is not necessary to evaluate greenhouses within these pages. However, there are one or two points that experience has proved worth mentioning.

It is accepted that wooden structures are warmer, quieter, easy to fix apparatus, linings to, etc. and look more natural. Complete panes of glass can be used, thus avoiding the need for clips and that awful line of dirt and algae which the overlap of small panes of glass encourages. The disadvantage of a wooden greenhouse is maintenance, but today's modern timber preservations have made the task a great deal easier. Many of the pieces of apparatus offered as optional extras are in fact essential. The option of a sliding door or normal hinged door is perhaps not thought about seriously but imagine a windy day when the door could be blown off its hinges, or the time when hands are full when a sliding door can be manoeuvered with elbows, hips or even a finger if necessary.

Automatic opening and closing vents are vitally important in at least one window – how many times has the greenhouse been closed up whilst the owner is at work or away and on return has found all the plants showing signs of overheating or drying out? On all but the coldest, frosty winter's day one window should be kept slightly ajar. To install a louvre window at ground or near ground level would create a complete circuit of air movement which is very important to the *Pelargonium*. Roller blinds or lengths of plastic shade netting (fixed with curtain wires and hooks or drawing-pins) are more satisfactory than the summer permanency of painted on shading. Corrugated PVC roofing sheets of the small profile type laid on strong supports will allow light to show through shelves and staging and, therefore, floor-space may be used to its full capacity.

Capillary watering systems are not truly recommended for *Pelargonium*, mainly due to the way the *Pelargonium* takes up water very quickly and in often large amounts after a drought period in the wild. In a greenhouse situation the constant supply available is still taken up even though not required and this could cause a complete breakdown of the plant if water is constantly on offer, as is the case with capillary systems. Another point is that no two plants need the same amount of water at any given time. If the owner is to be away for more than a few days during the summer, then perhaps this system will save the plants, that is if there is not a willing neighbour acting as caretaker/waterer. Still on the subject of watering, there are moisture meters which will, when the probe is inserted in the compost, give a fairly reliable reading of the soil's moisture content. If plants are being lost through over-watering and the owner is one of those 'compulsive waterers' that are, it is feared, of the majority, then the purchase of a moisture meter will repay itself a thousand fold.

The use of negative ionisers for greenhouses is becoming popular. This piece of electrical equipment is quite expensive to purchase and, of course, as with all electrical equipment, must be fitted safely by a qualified electrician. Even though the use of negative ionisers is fairly new to the greenhouse it has some advantages particularly during the winter months when the air and atmosphere are often dull and still because it will create a buoyant, clear and dust-free, alpine-like situation in the greenhouse.

Polythene lining of the greenhouse for the winter is a subject causing a good deal of argument. The practical use to conserve heat is admirable but the amount of light that sheeting, even the cleanest, blocks out is more of a loss than the heat that escapes. Providing the house is frost-free, light is far more important than warmth to *Pelargonium* during a British winter. The condensation caused by polythene is not good in a greenhouse housing *Pelargonium* and perhaps erecting a framework covered in polythene on the outside of the greenhouse roof would eliminate the drips and enclosed damp atmosphere. High winds could damage any outside polythene structure unless ample strength has been incorporated in the design of the protection.

There is no getting away from the fact that to keep *Pelargonium, Monsonia* and *Sarcocaulon* through the winter in Britain, the greenhouse will have to be heated to an absolute minimum of 40°F and preferably higher especially if it is hoped the plants will continue some flowering during the winter. It is generally accepted that an electric fan-heater, on a thermostat, is best but has the drawback of being expensive to run and there is always the possibility of power cuts. Oil and paraffin-type heaters need to be filled regularly and can run out – some *Pelargonium* seem to experience stress with this type of heater. Perhaps the most reliable form of heating is natural gas – it is not the most expensive fuel and gas heaters can be fitted with a thermostat. It could be a problem to serve a greenhouse if it is situated at the end of the garden and the apparatus must be installed by a qualified fitter.

Bottled gas in the form of propane is more convenient (although not as cheap as natural gas), as it is in portable containers which can be stored outside the greenhouse in cold weather. Although propane can be purchased in large cylinders the size accommodated in most gardens would be of about 19 kg capacity and so the possibility of running out is a problem even with the 'pig-tail' system of multi-connecting cylinders. The use of a solid fuel system is not in favour with amateurs both for cost reasons and the work involved with fuelling and cleaning such a system continually. As most heating systems have drawbacks, it could be that two of these systems should be used in conjunction.

The author's greenhouse is basically heated with a propane gas thermostatically controlled heater plus a back-up of an electric fan heater situated at the opposite end of the greenhouse from the gas heater and set at a minimal temperature to allow for it to cut in when the temperature plummets below 36/38°F during really icy winter weather or when the gas cylinder unexpectedly runs out. This system works well and fairly economically. Economics do have to be considered but often the grower does not balance running costs against the cost of replacing lost plants or the awful trauma of seeing a greenhouse full of frozen specimens, or calculate the material costs against the pleasure and pride of growing a *Pelargonium* collection.

Heated propagators come in many sizes and forms. For propagating during the colder months they are a must but it is not necessary to purchase a complete unit comprising of a seed tray with the heating element as an integral part and a see-through cover if *Pelargonium* are the only plants to use the equipment. Most growers will have these as stock items so all you need is the heat giving base, which should be with a thermostat if possible, then the pots and trays are stood on the base. If electricity is not available there are on the market propagators heated by small paraffin burners.

Any form of lighting in the greenhouse will be an advantage, whether it be electric from the mains or from a battery system. Lighting, usually recommended in the form of special horticultural lamps, can be a useful aid if plants are to continue to grow, flower and produce new growth during the cold, dark parts of the year and advice on this type of lighting may be obtained from the Electricity Council – in most cases lighting is used as a convenience so that one may enter or work in the greenhouse during the hours of darkness. Another piece of electrical equipment which to the *Pelargonium* grower should be a 'must' is an air circulating fan. This will keep the greenhouse at a constant and even temperature and also keep the air buoyant and fresh both in the summer and winter months. During the latter it is most important to keep the air circulating because most greenhouse windows seem, unwisely, to be kept tightly shut during this period of the year. Good air circulation will help reduce disease.

PROPAGATION

Seed

Seed is the easiest way to produce more species plants and will almost without exception come true to type.

Sow seed as freshly gathered as possible with the casing and awn still attached. A seedling compost mixture should be prepared first using a 50:50 mixture of proprietary seed compost and either horticultural sand, perlite or vermiculite. Mix and aerate with the hands before placing in small pots. Place in a bath of tepid water until the compost surface is wet, then drain well. Write a label with the name and other details needed then push the pointed end of the seed into the compost so that it is about one-third buried – leave the awn's coiling and uncoiling mechanism to dictate the actual depth of the seed depending on the moisture and heat conditions available. Cover very lightly with a sprinkling of dried sand. Leave in a shaded place, only watering when necessary. With old seed it might be best if the awn and casing are removed in case the outer casing is tough and hard and water will not penetrate easily. It is practical to chip or nick old seed – this is best done, carefully, at the pointed end, removing a fragment of the outer shell so that moisture may enter. Sow these seeds in the same way but push the seed down into the compost until just below the surface. *Pelargonium* seeds can begin to show signs of germination in twenty-four hours or may take up to a couple of years but don't be alarmed, the average maximum time is six to eight weeks under normal conditions. Do not discard old seed before trying to germinate them because *Pelargonium* seed can stay viable for a number of years and then germinate satisfactorily.

The two cotyledons that appear are usually oval to roundish and light to medium green in colour with some having a red tint on the underside – the seedling's stem is normally green but may be brownish. As soon as the next set of leaves grows, pot up, holding only the leaves whilst this is being done. The potting mixture should be a 50:50 mixture of John Innes No. 1 and a drainage medium of one's choice, well mixed together making sure there are no lumps. Use small pots to start them off, repotting when required using a John Innes No. 2 compost and the same proportion of drainage additive. Remember to take care with watering, place in the shade for a few days and make sure the plant is in an airy but not draughty situation. When the plantlet is ready for a normal sized pot, say four to five inches or larger (10 to 13 cm or more) the quantity of drainage material may be cut down if the species in question warrants this. Start applying the appropriate pot dressings and continue as stated in the previous instructions relating to potting and culture.

Cuttings

Before beginning this operation, gather together all the equipment and apparatus needed; a clean razor sharp knife, clean seedtrays or pots, labels and marker, sphagnum moss peat (well sieved) and horticultural sand or perlite or vermiculite, not forgetting a healthy stock plant or the plant material from which the cutting is to be taken. The working area should also be tidy and clean, washing down the bench if necessary with a Jeyes Fluid solution to the manufacturer's specification.

First mix up the cutting medium carefully, aerating it thoroughly, fill the trays or pots just to the brim then press down lightly with the base of a clean pot. Place in a water-bath until the warmed water just breaks the surface of the compost; leave to drain well. If there is a risk, due to the time of year and conditions, that the cuttings might remain too damp, a fungicide can be added to the water at this stage. Another aid would be to sprinkle a layer of dried horticultural sand over the whole of the compost before inserting the cuttings – this will fall into the cavity creating additional drainage at the base of the cutting, perhaps acting as a stimulant to rooting as well as keeping the surface of the compost drier. If the cuttings are taken in the winter, it has been proved effective to place a layer of perlite at the bottom of the tray or pot, as this will keep the compost slightly warmer and more well drained.

The method of actually taking species cuttings depends on the variety of species. For the beginner, and as a general rule, the following simple technique is advised. Having chosen a healthy plant or piece and written a label with the data required, using a clean sharp knife remove the portion of the stem at just above a leaf node on the plant by cutting with a slanted cut, where possible with the slant facing away from the top of the stem and any new growing bud that might be evident. Always bear in mind the shape of the stock or mother plant when removing material and trim back any parts to neaten and improve the plant's shape at this stage. Trim the removed portion to about 2-2.5 in. (5 – 7 cm). Make a straight cut across the bottom of a node, trim off any flowers or flower buds, any dead stipules or any stipule that may be underneath the compost. Trim off the excess leaves by breaking, using an upward pressure at the leaf and stem junction – this will discourage any tearing of stem tissue – or cut off the leaf stem (petiole) with a knife about half-way along, leaving the remaining piece of petiole to dry and drop away naturally. Leave two leaves at the top of the cutting.

The easiest cuttings will be from the tip of the shoot, at the growing point – success will be achieved by using pieces from further down the stem, but ensure that the stem is not too woody and old or the cutting will not root. Often cuttings taken at many places down the stem will create a problem due to the leaves being much larger than those at the tip. It will do no harm, in fact it could be advantageous, to cut these large leaves in half, only leaving enough to maintain the plant's food manufacturing processess – this practice will ensure that no leaves are touching each other and will allow air to circulate better as well as reducing the area from which the plant could transpire.

When the cutting is prepared, push carefully but firmly into the cutting mix which should be firm but soft enough for the stem to enter without the need to use any form of dibber to make a pilot hole. By allowing the cutting to make its own hole the possibility of an air space at the base of the cutting is eliminated. It is really a matter of preference whether one cutting is placed in a small pot, five or six round the edge of a larger pot or a batch in a seed tray. There is some proof that, within reason, the more cuttings placed together the better they will root. Only use a hormone-rooting preparation if it is felt desirable, normally *Pelargonium* species will not need this type of encouragement and it is possible that fungal and other diseases might be transmitted with the constant dipping of plant material into the preparation.

Much work has recently been done to find out if the *Pelargonium* and some other plants respond to the use of certain vitamins during the propagation process and results are encouraging, but only a handful of amateurs are using vitamins to assist with the rooting of their plants. It must be said that as *Pelargonium* are easy to root anyway it is not as vital to reduce the rooting time and increase the percentage of 'take' for the amateur as it is for the professional grower. However, in the case of some species who are notoriously difficult to root, a simple mixture of vitamin C (ascorbic acid) in the formula of 1 x 250 mg tablet (available from any chemist), crushed and dissolved in 4 tablespoons (60 ml) of warm water, wherein the prepared cuttings are allowed to stand for thirty seconds before being placed in the compost, is suggested.

Finally, place the tray or pot in a shaded part of the greenhouse for a few days, removing them to a place in more light but out of any direct sun. Only water if dried out and use the water-bath method of soaking and draining in preference to a watering can. If it is necessary to use bottom heat to encourage rooting the soil temperature should be about 55–60°F (12–15°C) – the use of a heated propagator, without the cover, will be needed during the winter and late autumn months. Under perfect conditions roots will start to appear from about ten to twelve days with the easier species, others will take much, much longer, but as long as the material looks healthy and feels firm when gently touched just be patient!

The question of how to tell whether a cutting is rooted is quite common. It is certainly a case of experience when one can look at a batch of cuttings still in the tray and be able to determine which have begun to root. Look for any strong signs of new growth, a springiness when the foliage is flipped and a general glow of healthiness. The only other way is to very carefully remove, with a two pronged table fork or similar, a couple of cuttings. If they have rooted take note and observe the state and condition of the cutting and soon it will be easy to ascertain if rooting has occurred in the future.

It is better if cuttings are potted on as soon as roots develop as this way the new plant will soon adapt to the compost, light and other aspects of being potted on and left to fend for itself. Use a good quality loam-based compost with added drainage mediums as advised earlier and leave in the shade for a while. Where possible hold the young plants by the leaves rather than the stem – they will nearly always grow a new leaf if one becomes damaged but to grow a new stem is a different story. The peat pots and pellets discussed in Chapter 1 may be used but with even more care regarding watering – again the possibility of being able to see the roots emerge is an advantage for the beginner.

This portion on cuttings will be used during the section on cultivars when this simple cutting method will be enlarged upon and perhaps the reader would like to try or experiment with the many different techniques which will be described later as alternative or additional ways of taking cuttings of *Pelargonium* species.

Root Cuttings

Root cuttings are possible from those species with large fibrous or stoloniferous roots. The basic procedure is explained in the relevant *Geranium* section but if *Pelargonium* root cuttings should be attempted in the winter or spring it will be necessary to provide a form of bottom heat in the greenhouse. When they are ready for the potting up stage, use a compost mix of half John Innes No. 2 and half the chosen drainage material – in subsequent potting the drainage material may be reduced or adjusted depending on the species in question. Keep the plants in the greenhouse and in a place sheltered from the sun for a while.

The breaking off of small tubers from the roots of tuberous rooted forms and then potting on straight away into individual pots of a well drained compost will multiply the tuberous rooted ones. Where these tubers are large enough, it is possible to cut the tubers almost right through, usually in two sections, each with an 'eye' evident, and leave them to 'chit' in a warm place out of direct sunlight, just like starting off seed potatoes. Make sure the cut edges are dusted with Captan or a sulphur powder to dry them off and guard against fungal spores entering the cuts. When the shoots are growing well, break the tuber away from the remaining piece, dust again and then plant individually into a very well-drained compost – do not water for two or three days to give the new wound a chance to begin to heal.

Layering

This method is not too often used with *Pelargonium* due to their being so ready to multiply with conventional cuttings but if the same principle is followed as with *Geranium* it is another way in which to propagate. Use pots filled with a gritty compost mix and place near the proposed mother plant or if the plant is in the garden for the summer months just sink the pot nearby. When rooting is evident, pot up in the usual manner.

Dividing

This is not a very suitable method for the species, again most will propagate using easier ways. It would be a pity to uproot a good plant to divide it – anyway the two or more extra pieces would take a lot of care and a long time to regain the original good shape required. As an absolute last resort it is possible with the fibrous ones. Dividing those species that produce off-sets on or just below the soil surface is a simple matter of carefully easing away the small off-set from the parent plant and potting up into the usual potting medium.

HYBRIDISING

Species *Pelargonium* hybridise in the wild on occasions and this has led to some of the confusion regarding naming some of the species. With some there may be a difference between plants having the same name or those looking alike but having a different name – these differences may be slight or extreme. This is not always due to natural crossings but often due to geographical conditions and situations making it necessary for the same species to adjust and equip themselves in differing ways, to cope with the conditions available in a particular area. This change can be visually obvious in the form, size or colour of a flower or leaf.

There have in the past been only a handful of people who have seriously been involved with crossing the species. It is a matter of choice as to whether the true species can be improved upon. There have been interesting and delightful developments over the years particularly with the Scented-leaved forms.

Obviously, with the promise of near black or sulphur yellow flowers with a beautiful perfume, there is room for hybridising within the species, it cannot be emphasised too strongly that strict, detailed records of all crosses must be kept and if a revolutionary new cultivar is developed it should be registered with the National Society and with the International Registration Authority in Australia.

The basic rules of hybridising are mentioned in the *Geranium* section and the cultivar section. Cross-pollinating in a greenhouse is a little easier than working out in the open garden but the bagging and very careful watch for receptiveness is still crucial.

PESTS, DISEASES AND DISORDERS

Whitefly would seem to be the most persistent greenhouse pest taking up residence on the underside of foliage. Undetected the female will lay about one hundred and fifty eggs at a time which will hatch in approximately ten days and the young soon reach the time when they can repeat the process producing, in turn, millions of tiny dusty white, moth-like insects with four wings. Glasshouse whitefly (*Trialeurodes vaporariorum*) have a piercing mouth and suck sap from the plant, preferring some types and even varieties more than others, and they can transmit disease as well as weaken the plant. They do not often fly in swarms until the plant is touched and then a bad infestation can be seen rising like a pall of white smoke from the plant.

Most chemicals available to the amateur are unable to dispense totally with the problem due to the whitefly having built up a strong resistance to many of the proprietary preparations. The old fashioned remedy of a liquid soap spray has been used with some advantageous results. As long as it is not a highly detergent based soap, washing-up liquid, mixed one-quarter teaspoon to two pints of water (1¼ ml to 1¼ l) and sprayed above and beneath the foliage, will be reasonably successful. This will only polish off the adults so the process should be repeated at least once a week to catch each hatching, until the problem is eradicated. This soap spray will also clean off the unsightly and potentially dangerous fungal growth seen as black deposits on the foliage. It is caused by the whitefly excreting honeydew and the resulting sooty fungus will impair plant vigour as well as make the plant very unsightly, so it must be removed.

The whitefly's attraction to yellow has prompted research and perhaps a homemade sticky concoction spread onto a yellow painted strip of wood to trap the insect is worth trying but do hang these strips away from the normal thoroughfare of the greenhouse or conservatory, for obvious reasons! As with most pests, the real answer is constant searching and careful inspection of one's plants, if this is done regularly the few pests present can be squashed between finger and thumb. A deterrent would be not growing plants which attract whitefly such as tomatoes, cucumbers and fuchsias.

Greenfly and most other aphids will attack the youngest shoots and growth on a plant as well as some semi-mature material and blooms. They also suck the sap, creating the same problems as whitefly. If the new shoots on plants appear crinkled and distorted, checking for greenfly will most likely reveal a colony. The soft bodies cast their skins at times and these can cause an unsightly problem as well as being a breeding ground for fungal diseases if in a wet condition. They will also give off a honeydew deposit. The treatment for greenfly can be as for whitefly although some chemical and organic preparations do seem to be more successful in dealing with the greenfly. The use of predators in the greenhouse, to control all types of aphids, including whitefly, is becoming popular for the amateur grower. Some already dwell in the garden, such as ladybirds, hoverflies, lace-wings, various beetles, spiders and centipedes – some may be purchased from firms specialising in the biological control field.

There are root aphids too, as well as root mealy bugs and it does mean using a proprietary preparation, watered onto the soil to dispose of these tiny, powdery coated, white insects which will devour small roots. If a loam-based compost is used with adequate drainage it is unlikely that a problem will develop. They can be detected often on the inside of the pot wall when a plant is removed from the pot and seen amongst the root system also. The damage can be quite severe to the point of the first signs of their presence being the wilting of top-growth and when the pot is tipped out the roots may be severed from the plant. This also applies to sciarids (*Pseudococcus spp*) which are small black flies that scurry over or fly over the compost surface. The larvae are thread-like maggots living in the soil, eating old vegetation and disturbing the root system. If this state is reached the best way is to wash all the compost from the roots in a solution of an insecticide specified to treat aphids, then pot up with new, good quality compost.

Mealy bug (*Pseudococcus*) can attack the stems of succulent species and are small whitish grey insects about 0.25 in. long (5 mm). They will secrete a fluffy white covering over colonies and eggs. It is unlikely that anything but a matchstick tip soaked in methylated spirit and dabbed onto the creature will be required if the mealy bugs are treated on sight.

The tuberous-rooted species and others that die down for periods can be attacked by root eel worm (*Nematode*), this small pest is not common but will destroy or distort root systems. If suspected or discovered the compost should be replaced and the roots washed in an insecticide solution, or a granular, basic garden preparation for root pests used.

Caterpillars will normally be found outside in the summer months and sometimes find their way into the greenhouse. Their presence is detected by large holes in the foliage and sometimes in the new flower buds. Usually only one or two will be discovered if an inspection programme of plants is carried out – these can be picked off the plants and disposed of. If this is not possible or they cannot be seen during the day which is quite common, then a night-time caterpillar hunt will be worthwhile. Failing this there are both organic and chemical preparations available.

Botrytis cinerea or grey mould is a fungal disease usually found at a time when cold, damp conditions are present. Falling petals and other plant debris are marvellous places for the spores to develop. New cuttings and seedlings are mostly at risk and also sick or very young plants. At times when a mature plant has been damaged *Botrytis* will attack. If a large plant is affected in part, cut off the diseased portion and dust the wound on the plant with a fungicide powder – it may be possible to make a cutting from the removed piece if the mould is not too severe but on no account use any piece of affected material. It is very important to sterilise the knife used before any other cuts are made. If the problem is widespread spray with Captan or copper fungicide as directed on the container but if the greenhouse and plants are kept clean and warm and ventilation is good, *Botrytis* should not be a continual problem.

If seedlings damp off this could be caused by various soil-borne and water-borne fungi. Do not allow seedlings to become cold or overcrowded and ensure that all compost is sterilised or freshly purchased and all pots and other apparatus clean. A spray with a warm solution of a copper fungicide or cheshunt compound at the pricking on stage will help avert this tiresome problem.

Occasionally black-leg may be a problem but as this usually affects cultivars rather than Species it will be dealt with later.

In some species a reddening of the foliage may occur, this is usually due to the plant being in a draught or in too cold a greenhouse, or a sudden drop in general temperature. Sometimes, in the late spring when the sun begins to strengthen and the new growth is still delicate this sudden strong sunlight will cause this disorder. The affected leaves will need to be removed, they will not turn green again but slowly deepen in colour, die and wither. It may be that some kind of deficiency has arisen due to an imbalance of chemicals in the compost.

This can cause the foliage to take on different colours in various parts but if plants are treated to a good quality compost, repotted and fed correctly this should not be a problem.

As has been said before, most pests and diseases will only occur if a plant is sickly; if hygiene techniques are abandoned or if unclean stock has been introduced to the collection before any check made. Where possible do make use of the natural deterrents and remedies before resorting to chemical preparations, and do make sure the instructions on the container are at all times legible, understood and followed precisely.

EXHIBITING

The showing of Species and the Scented-leaved types is becoming more popular and the Joint National Societies have compiled a judging points table which suggests a standard maximum percentage of points which may be used in the awarding of prizes as follows: 50 per cent for cultural quality, 30 per cent for foliage, 10 per cent each for flowers and for staging and display. So it can be seen that the overall size, shape, condition, cleanliness, proportion, quality and in some cases difficulty of cultivation can take half the points and the shape, size, number, spacing, condition, cleanliness and scent if applicable are more vital than the blooms and general staging. Not that the last two are any more unimportant, many an exhibit has been knocked from its pedestal with an unclean plant, dirty pot or no blooms. Inspect the potential show specimens carefully, paying attention to the finer points, including a check as to whether a plant is taking any pests or diseases to the show.

Stopping and disbudding of Species and Scented-leaved will not be an important consideration – it is not recommended with most of the Species but with some of the larger Scented-leaved stopping or pinching out will encourage a good shaped and well filled-out plant as some certainly do become straggly. This operation should begin as soon as the plant is potted into its first pot and continued up to about ten to twelve weeks before the show season to give the plant a chance of plentiful flowers and the new tips to grow evenly.

The Species should follow the form of those growing in the wild wherever possible and not be encouraged to grow huge and 'show-like' if this is not natural. If plants are setting or have set seed, this, it is thought by many, should be left intact and form part of the exhibit.

A top dressing of a natural material such as gravel or sand should be provided but do not exhibit a plant which still has the grass growing in the pot (as was suggested is possible to give the plant a natural habitat effect in the greenhouse).

Individuals, clubs or societies may be asked to show displays of Scented and Species *Pelargonium*, not necessarily for exhibiting purposes but usually as a general interest for the public. This is a chance not to be missed by the enthusiast. It is not necessary to present perfectly grown large, show-type plants because two or three of the same sort, placed together to make a group, with others showing lots of bloom and good foliage will give an excellent overall picture. The ideal is to have a box-like structure made in which the plants stand still in their pots. Filling the gaps with clean sphagnum moss or peat or bark will make the display look natural. The addition of a map to show where the species originate, good, well-written name labels and some information leaflets will encourage a few more to this fascinating section of the Fancy.

PELARGONIUM SPECIES PLANT LIST

Section *Campylia*

P. ovale: Short, shrubby, few branches. Leaves usually oval, grey in colour and covered in tiny grey hairs. Flowers, palest to dark pink with markings of dark pink to maroon on upper petals. From the Cape Province. 4–6 in. (10–15 cm). E
P. violarium: Leaves lanceolate with some notchings at margin, sage-green with fine hairs covering. Shrubby, short stemmed with branches spreading. Flowers, upper petals purple and maroon with large dark red markings, lower petals whitish-mauve. From Cape Province in sandy soils. 4–6 in. (10–15 cm). E but a beginner should try after some experience.

Section *Ciconium*

P. inquinans: Tall branching, soft woody plant. Rounded to heart-shaped foliage, mid-green. Flowers usually blood red to vermilion. Eastern Cape region. 18–36 in. (45–100 cm). B

P. acetosum: Slender stems, leaves glabrous, oval with base broader, some marginal waving and notching. Blooms of a definite two upper petal, three lower petal formation. Salmon to coral, occasionally pale salmon in colour. Eastern Cape area. 10–18 in. (25–45 cm). B, with care

P. zonale: Erect or scrambling through scrub and shrub land. Leaves nearly orbicular with narrow, heavy zone on mid-green, shiny leaves. Blooms are pink but red and white are common in the wild. Most places in Southern Africa and along coastlines. Normally 36–40 in. (1–1.25 m). B

Section *Cortusina*

P. echinatum: Shrublet with spiny, succulent, grey-green stems. Leaves heart shaped and heavily veined, green to grey-green. Flowers are pure white with heart-shaped red blotches on top two petals giving it the nick name 'Sweetheart Geranium'. There is a pink-flowered form also. From Namaqualand in dry, stony places. B with care

P. reniforme: Kidney-shaped leaves, slightly hairy and soft. Pink to magenta, medium sized blooms. Tends to straggle along the ground. Eastern Cape in grassland and dry plains. 10–16 in. (25–40 cm). B

Section *Dibrachya*

P. peltatum: The ancestor of the Ivy-leaved *Pelargonium*. Leaves are of an ivy leaf formation. Glossy, strong, fleshy and aromatic. Mid-green in colour. A climbing or scrambling plant, as all in this section are. Flowers are pale lilac to deep mauve and occasionally white, all with some markings in top petals. Lives in grass amongst rocks and shrubs in Cape Province. Stems up to 6 ft (2 m) long. B

Section *Eumorpha*

P. alchemilloides: Shrubby, slender plants with trailing tendency. Palm-shaped leaves with often a deep zone and green hairs evident. Flowers smallish, either mauve-pink or cream to yellow. Grows through shrubs in most of Africa. 8–15 in. (20–40 cm). B

P. grandiflorum: Palmate leaves of grey-green, sometimes zoned. Aromatic. Large pale pink or mauve blooms, blotched on top two petals also veined. South Western and Western Cape in mountainous regions. Up to 18 in. (45 cm). B

Section *Glaucophyllum*

P. glaucum: Small growing sub-shrub. Lanceolate, glaucous foliage with entire margins. Blooms medium sized, yellow with maroon markings on top two petals and veining of same colour in lower. From small area of Southern Africa. Unusual and lovely. 8–10 in. (20–5 cm). E

P. laevigatum: Straggling half-shrub, woody on old stem parts. Leaves composed of three well-cut leaflets, mid-green and greyish. Smooth. Petals usually mauve-pink with markings and veinings of purple. Small blooms but many and for a long period. Often deciduous. Cape Province. Height variable from a few inches to 24 in. (60 cm). E

Section *Hoarea*

P. longifolium: Tuberous rooted geophyte. A species very variable in leaf form from long entire leaves to deeply cut and divided even on the same plant. Mid-green leaves with long hairs over all parts of the plant. Flowers white with dark purple markings and veins. 10 in. (25 cm). E

P. rapaceum: Tuberous rooted. Foliage carrot-like and very hairy, grey-green in colour. Blooms smallish, many to an umbel, pale to dark coral sometimes yellow. The lower petals remain folded, looking as one. Mainly South West Cape in sandy soil and stony slopes. Fairly ground hugging with flower stems up to 10 in. (25 cm) high. E

Section *Isopetalum*

P. cotyledonis: The only plant in this section heralds from Saint Helena. Thick, succulent branches and stem. Leaves are shaped like miniature water-lilies and are very veined, shiny and mid-green in colour. In the autumn the foliage takes on russet hues before defoliation. The flowers are white and small. Known as 'The Old Man Geranium'. Up to 12 in. in height (30 cm). E

Section *Jenkinsonia*

P. praemorsum: Shrubby, woody, slender stems with unusual zig-zag formation. Small green leaves, slightly aromatic, kidney and palmate shaped and deeply divided. Blooms large for the plant size, four petals of yellow with dark red veinings, the upper two more marked and larger. Found in Namaqualand. 30 in. (80 cm) at most. B

P. tetragonum: Four sided grey-green, succulent stems. Leaves, palmate, crenate and lobed. Some hairs present on leaf margins. Flowers of four petals, normally pink but may be cream all with deep red veinings. The upper two petals are larger and more marked. Southern Africa. Length of straggly branches up to 6 ft (2 m). B

Section *Ligularia*

P. crassipes: Small growing shrublet with leaf petioles persistent as long, dry spikes. Leaves carrot-like and green with coarse hairs. Small, pretty pink blooms. Found in an area of Namaqualand. 6–8 in. (15–18 cm). E

P. hirtum: Low-growing and bushy with dense carrot-like leaves of grey-green and covered in long hairs. Flowers are rounded, deep pink and dark purple spots on top two petals. Cape Peninsula in sandy soils and rocky ledges. 6 in. (15 cm). B, with care, easier from seed.

Section *Myrrhidium*

P. myrrhifolium: More than one type are found. Generally a shrubby plant with green leaves, deeply cut and divided, hairs cover stems. Flowers, white, pale mauve or pink of four petals. Markings of red on top petals. A tap-rooted system evident. South Western Cape. 12–16 in. (30–40 cm). B

P. urbanum: Tuberous rooted type. Very segmented leaves of green, lobed and cut with hairs and longer hairs on petioles and stems. Flowers yellow or pale pink with scarlet veins on top two petals. Top petals very broad, bottom petals very narrow. Found in sandy places and coastal limestone ridges in South West of Cape Province. Rambling to 12 in. (30 cm). E

Section *Otidia*

P. carnosum. A succulent and branching plant with thick stems. Leaves on long mid-ribs are pinnately divided and fleshy. Small blooms of white or cream with some markings. From the Southern parts of South Africa, the Western Cape and dry parts in the Eastern Province. 14–36 in. (35–100 cm). B

P. crithmifolium: Very similar to above but with finer leaves. Blooms on long stems which remain for over a year and dry on the plant. White blooms with reddish markings. Grows in similar places to *P. carnosum*. 18 in. (45 cm) but normally shorter. B

Section *Pelargonium*

P. cordifolium. Leaf heart-shaped, sometimes concave, green with hairs dense on the underside of foliage giving a grey reverse to the leaves. Many flowers on the umbel of rosy-carmine flowers. The larger upper petals streaked with purple. Enjoys the Southern and Eastern Cape coastal areas. 36 in. at most (1 m). B

Section *Peristera*

P. australe: Many forms found in the wild. Usually low growing and spreading. Leaves cordate, lobed, large and grey-green with some hair covering. Autumn hues in season. Whitish flowers with dark pink or purple veins. From Africa, Australia, Tasmania and most islands in that region where each place seems to have its own type. Height from a few inches to 16 in. (1–40 cm). B
P. chamaedryfolium: Treated as a annual in the greenhouse. Seeds well. Leaves somewhat heart-shaped and lobed, green with the underside darker, sometimes russet. Flowers small but numerous of rich magenta. Southern Africa. Maximum height 12 in. (30 cm). B

Section *Polyactium*

P. gibbosum: 'The Gouty Geranium' due to the swollen nodes. Half succulent foliage of grey with a greyer overtone or bloom. The leaves seem to be of three divisions and the flowers are sulphur yellow, sweetly scented at nightfall. South West of Southern Africa, frequently in sandy, rocky coast lands. Scrambling to many inches. B
P. fulgidum: Partially responsible for some hybrids of today. Half succulent shrub. Leaves green with a silky overlay, pinnately and deeply cut. At the most 36 in. (1m). B, with care.

Section *Seymouria*

P. asarifolium: Meaning like the arum leaf. Hairy petioles bear the green, shiny, round leaves, these hairs densely cover the underside of the leaf making it silver in colour. The flowers are tiny, with only two petals, which are darkest red to purple. A geophyte. Found in a small area of South Western Cape in stony places. 3 in. (8 cm). E

Pelargonium **Primary Hybrids**

P. x ardens: Red flowered. E
P. x glaucifolium: Dark purple, almost black with thin yellow margin to petals. E
P. x schottii: Purplish red flowers. E

THE SCENTED SPECIES AND HYBRIDS

Usually known as 'Scented geraniums'. By this point of the book it is hoped that the reader will know this title is incorrect, firstly because they are not a *Geranium* and secondly because the usual definition of 'scented' refers to the scent of the flower, whereas in this instance it refers to the leaves. It has been accepted that 'scent' is to flowers whilst 'aromatic' is to foliage. Strictly speaking then, they should be known as 'Aromatic-Foliaged Pelargoniums'. One can see why they are not! So, if in this chapter they are referred to as 'Scented *Pelargonium*' and 'Scented geraniums' it is because these are the titles used by both hardened enthusiasts and interested members of the general public.

The Scented-leaved types are Species or derived from Species and have an in-built defence against animals or insects by producing an aromatic oil through glandular hairs on the foliage and stems, or to some insects an attraction to visit the bloom and pollinate. This scent is expelled when the plant is bruised, touched or moved. To us, most of the scents are very pleasant but to some predators they are not so and in fact they can render the leaves repulsive or even indigestible. There is some belief that the oil gives off a vapour which can protect the plant in harsh sun.

The first Scented-leaved species to arrive in England is thought to have been *P. capitatum*, the 'Rose-scented Geranium', during the seventeenth century. Today there is a notable industry extracting the oil from plants such as *P. capitatum* and *P. graveolens* to use in the perfume and soap trade and for the popular essential oils that are used in the making of pot-pourri and air purifiers, etc. 'Rose geranium' oil has as its two main constituents *geraniol* and *citronellol* and these are located in the green parts of the plant. There are many scents or aromas which basically can be grouped into the following general classification: Fruit, Spice, Flower, Aromatic, Pungent or Herb. At the end of this section a list of varieties with their scents will be found.

Roots, Foliage and Blooms

Roots

Most of the Scented-leaved types have a strong fibrous-root system. In some the roots are plentiful and will soon fill the area allotted to them – on the whole they show no problems with potting and re-potting.

Foliage

It is impossible to generalise on the shape, texture, size and composition of the leaves. Some have very tiny, almost round leaves, some the size of a man's hand and in each of these there are leaves which are either divided or cut. The surfaces can be shiny and hard, smooth or crinkled or soft and hairy, in colours ranging from light to dark green, grey or nearly brown.

Because both species and hybrids seem to cross and sport easily, many differing forms and scents are being introduced together with pretty variegated forms. The height and width of these specimens ranges from two to three inches (5–8 cm) to over 6 ft (2 m) or more if allowed. Some form neat hummocks, some scramble, some grow into tree-like proportions and these points should be noted when growing the plants 'in captivity'. For example, the question of whether they are to be grown and to live in a greenhouse or conservatory with ample top light or on a small, narrow windowsill where space is of a premium should be considered.

Blooms

It must be admitted that these are generally small to medium in size and whereas most have pale to deep mauve petals with darker markings there are some with white, pinkish/mauve and deep pink flowers. They vary in size from 0.125 in. across

Leaf shapes of scented types.

to 1.5 in. (2 mm–4 cm) with the typical flower structure of five petals, unevenly spaced, as a rule, with two above and three on the bottom. Because many of the flowers are small does not mean they go unnoticed – the opposite in fact, due to them holding their blooms to the fore as if they are showing off the least of their assets to the maximum. It is not difficult to encourage Scenteds to bloom throughout the year and a small rise in winter temperature will encourage this.

Cultivation

Scented 'geraniums' are the ideal basis of a house-plant or *Pelargonium* collection. They are, with their diverse foliage shapes and colours and many differing scents when rubbed, a 'talking point' and a natural interest for the blind, disabled and house-bound gardener. Scented 'geraniums' need much light so a table, bench or windowsill of a height to catch as much light as possible is essential; even so, the plant should be turned frequently so that it may receive as much equal light around the plant as can be provided. Ideally, turn the plant each day when on a shelf or in a window area that does not receive top light or all-round light adequately.

Normal central heating in the house or a heated green-house, particularly if fan assisted, will be an ideal home for the winter months. In the summer they may be placed out of doors. Remember the Scented-leaved *Pelargonium* will not tolerate frost. They also hate damp, dark and dank conditions. The problems of growing on a windowsill are mainly two-fold. It is essential that the plants are not left between the glass and the curtains on a cold night because the cold cannot escape further than the curtained window space thus often meaning temperatures are at freezing point for some time before the curtains are pulled back in the morning – neither should they be subjected to magnified heat and sunlight on a hot sunny day, particularly if the foliage has been wetted. Because of the many statures, the restrictions of windowsill space need not be so great.

In a conservatory or greenhouse it is assumed there will be plenty of light and so the whole range of Scenteds may be grown. Ensure always that the air circulation is good – the door may be left open on all but the frostiest days but make sure that the premises are frost-free. When severe frosts are threatened it is a good plan to ensure the glass walls do not allow frost to enter near the plants – a good method of keeping the plants safer during such spells is to place loosely crumpled newspaper between the plants and the glass or cover the plants with two sheets of newspaper. This procedure, hopefully, will only have to be resorted to a couple of times each winter. It will not harm the plants if the papers are left on for a few days and nights because the plants, in that temperature, will not be thinking of growing. Adequate air circulation will help the cold air move around rather than stay in one area so causing problems. It may be that some shading will be needed for the plants during the height of summer.

The temperature throughout the year depends on whether the plants are to remain as a flowering feature for most of the year or just kept for the foliage. If they are to be encouraged to bloom in the winter then a temperature of 45°F (7°C), higher if wished, will have to be provided, otherwise a soil temperature of 40°F (5°C) or just above will be adequate to keep the plants safely through the winter and early spring. Do take note on intensely cold days because the temperature will have to be raised if an acute drop in the night temperature is forecast.

As well as in the home, 'Scented geraniums' can be planted

outside in their pots as soon as the frosts are over. Used in the herb garden they provide an interesting and useful addition. They can be planted in window-boxes or in beds either on their own or with other subjects as part of a bedding scheme. If the garden soil is on the rich side it will be best to leave them in the pots due to the fact that 'geraniums' have a tendency to develop large foliage and fewer flowers in very rich soil, coupled with the British summer usually promising plenty of rain, so encouraging plenty of lush growth at the expense of blooms. This system of sinking the pots also helps when it is time to move the plants indoors in the autumn.

When planting in window-boxes it will add to the attraction if they are placed so that when the window is open the plants can be easily gathered for cooking purposes or for the flower vase – the scent will be a bonus particularly on warm evenings.

Normal British summers are acceptable to the Scenteds although very wet weather may cause problems if the ground or pots are not well drained, so do follow the basic advice for the cultivation of Species *Pelargonium*. All spent and dying flowers or leaves should be removed immediately, not only will this action encourage the plant to flower continually and produce new growth, it will also guard against fungal diseases and pests foraging and living in the plant debris.

Indoor over-wintering of plants that have been outside will be necessary and they must be removed to a frost-free place before the first hard frost, which is often early in October. The plants may be large by this time – it could be a good plan to prune them so that they will be easier to cope with through the winter. The prunings may be used for cuttings if they are suitable or used in the kitchen or for drying, etc.

When lifted either stand the pots underneath the bench for a few days to recover or take them indoors into a room that does not experience a high temperature and a closed atmosphere during the winter. Place them later on a windowsill or in plenty of light. It is possible to keep the plants at a low but frost-free temperature, just to keep them 'ticking over', withholding water and feeding until early spring. They can be kept growing and flowering throughout the winter, in which case the temperature will have to be suitable and more water offered, but always keep watering to a minimum.

Taking cuttings at the lifting stage means that old plants can be discarded and the cuttings provided with acceptable conditions for rooting. To store plants is the least favourable method but if this is the choice then lift the plants from the garden bed carefully removing as much soil as possible. Trim off with a clean, razor-sharp knife all excess or damaged growth, and dust with a fungicide powder. Place the plants upright in a box and cover the roots with compost mixed with sand. Water sparingly with a fungicide solution, then place in a well-lit and frost-free place for the winter months, checking at times for disease and dead foliage, etc. The old method of hanging plants up in a shed or leaving them under the spare bed may sometimes work but the difficulty arises on deciding when to bring the plants out again to start growing normally. More often than not, shoots are evident – these will often dry off and the plant takes a long time to recover, if at all. By far the best way is to try and keep the plant growing slowly and steadily in its pot, trimming off growth and damaged leaves or stems as necessary, only watering as needed then placing the plant under a bench to await recovery before placing it on top of the bench or on the windowsill to be enjoyed the whole winter through. If artificial lighting is at hand it can be used to an advantage, making a badly lit area of a room suitable for *Pelargonium* growing as well as encouraging the plants to flower freely during the winter.

Basically *Pelargonium* only need a small amount of water, increasing when coming into bud and more at the flowering stage. Always under-water rather than over-water. Those situated in the garden will be receiving enough moisture apart from times of prolonged drought. Feed sparingly and often, half strength at every watering increasing to three-quarter strength at each water just as the first flower buds show. A general, well-balanced liquid feed or one that can be made up into a liquid is best.

Normally Scenteds will need potting up once a year once they have become mature. No harm will be done by depotting to keep the plant to a reasonable and practicable size. This is best done in the spring before most growth has begun to move. Remove the plant from its pot and carefully tease away the soil from the roots, trim back most of the larger roots, trying not to damage the small ones and repot into the original pot, water well and then leave to drain in a shaded place for a while. A good quality John Innes No. 2 with added drainage will be suitable.

When the plant becomes too large for its situation pruning may be undertaken. Use a clean, razor-sharp knife or if the stem is woody, a pair of good, clean, sharp secateurs. Do not be afraid of being too ruthless, cut to within two or three leaf nodes if drastic pruning is needed or just take off the ends of shoots if it is a trim that is required. Hard pruning but leaving some growth, will lead to the plant branching out more, rather than becoming straggly which is a fault of a number of Scenteds.

Some of the smaller-leaved types will lend themselves to being trained, and it can be fun creating a standard, column, a fan or some other more adventurous shape. The more vigorous of varieties can be encouraged to grow up against a wall in the greenhouse or conservatory or fall from hanging baskets or hanging pots.

Apart from growing the Scented varieties as decorative pot plants the leaves can be used fresh as a substitute for some herbs or as a flavouring in their own right. The leaves may be chopped and placed into the ice-cube tray to be stored in the freezer until needed or the leaves frozen whole in the ice-cube – they may then be dropped into a refreshing summer drink. In addition the leaves of some flower or fruit scented ones can be used in home made toiletries. Dried, the leaves will make up a useful addition or base for many pot-pourri recipes. Crystallising the leaves in sugar makes an attractive and unusual cake decoration.

RECIPES USING
SCENTED-LEAVED PELARGONIUMS

Note: It is important to ensure that all leaves and flowers are washed thoroughly before use and if any possible contact with horticultural chemicals is suspected please follow the advice and precautions on the label of the preparation regarding time lapse before harvesting, etc. Don't take chances!

'My Favourite Pot Pourri'

1 pint by volume (½ litre) dried rose petals, as fragrant as possible
½ pint by volume (⅓ litre) dried 'geranium' leaves, mixed if desired
½ pint by volume (⅓ litre) lavender flowers or a few drops of lavender oil
½ teaspoon (2½ ml) each of powdered cinnamon, mace and cloves
2 teaspoons (10 ml) orris root powder
A few drops of 'geranium' oil

Mix flowers, petals and leaves. Leave in the open for one day to ensure they are crisp. Mix in the remaining ingredients, store in an air-tight container for three weeks, mixing well once a week. When required, set out into pretty dishes, the remaining may be stored in an air-tight container for many months.

Rose 'geranium' Water

1 handful of scented 'geranium' leaves of the rose-scented type
1 pint (½ litre) water (clear rain-water if possible)

Simmer together for 15 minutes, steep for 2 hours, then strain. Larger quantities, perhaps at a time when the plants are being pruned or cut back, may be made and frozen until required in freezer bags. As an after shampoo rinse, run the water three or four times through the hair to give a pleasant fragrance and shine. The preparation may also be used as an astringent for the face and neck.

'After Gardening Hand Cream'

6 tablespoons (90 ml) of rose 'geranium' water as above
1 heaped teaspoon (5 ml) of borax
1 disc, approx. 25 grams of white beeswax, or well bleached beeswax
8 tablespoons (120 ml) white petroleum jelly or Vaseline
6–8 drops 'geranium' oil
A few drops of red or pink food colouring

Dissolve borax in the rose 'geranium' water. Melt wax in a saucepan over a very low heat and very carefully. Using a wooden spoon or spatula, add the water solution, stirring continually. When cool and thickening add oil together with the colouring to give a soft pink shade, and blend well. Pot, seal and label. This will make enough hand cream to fill a 1 lb (½ kilo) jar and put into small decorated pots will make nice gifts.

Lemon 'geranium' Cake

10 leaves from a lemon-scented variety, well washed and dried
2 tablespoons (30 ml) milk
2 eggs
4 ounces (100 grams) castor sugar
4 ounces (100 grams) 'tub' or whipped margarine

5 ounces (125 grams) self–raising flour
pinch salt
1 level teaspoon (5 ml) baking powder

Place leaves and flour in an airtight container for two hours, pressing the leaves and stirring occasionally. Remove leaves and arrange on the base of a greased cake tin. Place the rest of the ingredients into a large bowl and beat all together for two minutes, until mixture is smooth – do not over beat. If using an electric mixer only beat until all ingredients are incorporate and smooth. Place in cake tin on top of leaves and bake at Gas Mark 4 or 350°F (180°C) for twenty to twenty-five minutes or until cake begins to come away from the tin sides. Remove from the tin – the leaves may be removed or left on the base. Split the cake in half and fill with seedless jam.

Geranium Cooler

3 leaves from a rose-scented variety
3 leaves from a peppermint-scented variety
1 leaf of any scented variety to be used as a garnish or decoration
6 whole cloves
2 tea-bags
A very small pinch of salt, not enough to taste
Sweetening if desired

Pour ½ pint (¼ litre) boiling water over all the ingredients, cover and infuse for five minutes. Sweeten to taste, perhaps with honey. Strain, cool and chill, and serve over crushed ice with a leaf in each of the two glasses. Cool and refreshing for a hot summer's day.

Propagation

Seed
Scenteds and the hybrids can be grown from seed quite success-fully and normally come true to the parents' form, although they will be of an undetermined pedigree sometimes.

Nearly all Scented-leaved 'geraniums' set seed so there should be no problem acquiring them but it is a case of exchanging with other enthusiasts because these seeds are not catalogued very often, if at all, in the popular seed catalogues. Seed sowing should be carried out as previously advised and they will germinate quickly on the whole with quite a high germination result, with some varieties so high that sometimes it looks like mustard and cress in the pot! When seedlings are large enough to handle by the first two true leaves, pot them up singly into pots of John Innes No. 2 with added sand or grit or other drainage material. Most will require larger pots by the end of the first season. To keep them pot-bound usually causes straggly growth and reduces the scent of the leaves.

Cuttings

If no heat is available it is best to take cuttings during the summer months. Choose a healthy stock plant, write the label and then carry on as described in the Species sections on cultivation and propagation. Save any wholesome leaves that are discarded and dry them or chop and freeze for future use.

Layering

Layering is also possible but not often used for Scenteds, perhaps because as a rule they are grown as house plants and not left to scramble but trained and neatened to be accommodated in the home. Most are easy from normal cuttings so it is hardly worth the extra bother of layering but it may be achieved following earlier instructions.

Hybridising

Many new varieties have come into the hobby in recent times. The aim seems to be larger flowers, more flower colour variation and a stronger leaf aroma, good enough reasons indeed. Most Scented types find no difficulty in self-pollinating so if serious hybridising is to be carried out the use of bags and a constant checking of receptiveness is vital. Some hybridists are working on the crossing of Scenteds back to the more unusual and beautiful true Species *Pelargonium* with outstanding results – it will be interesting to see what the next few years achieve.

Pests, Diseases and Disorders

These will be found in the Species section

Exhibiting

The showing of Scenteds comes in the same classification for marking and staging as Species *Pelargonium*. Some shows have classes for Species and Scenteds in the one class, which is a pity because not only does it make it difficult for the judge it does not give the exhibitor or the plant a fair chance.

In the Floral Art classes the Scenteds can come into their own with the wonderful array of foliage shapes and colours on offer. Sprigs of the tiny leaved types and also the largest leaves will keep well in water and stand well in oasis or pin holders. The small leaves and flowers are excellent for the miniature and petite classes.

Choose blooms from the varieties whose single flowers do not shatter quickly, if in doubt try a few flower-heads in water a week or so before the show, just to make sure. A gentle spray over the whole arrangement just before placing it in its final staging position will keep it fresh for the duration of at least a three-day show and still give some pleasure when taken back home.

Scented-Leaved Plant List

There are many scents or fragrances which can be grouped into the following general classifications:

1.   Fruit scents, which include orange, lemon, strawberry, filbert nut, lime, citron, apple, and coconut.
2.   Flower scents, which include rose, lavender, and cedar.
3.   Spice scents, which include pepper, ginger, spice, nutmeg, and cinnamon.
4.   Herb scents, which include peppermint and peppermint/rose.
5.   Aromatic and Pungent, which are difficult to describe – some are pleasant, some are not.

P. abrotanifolium: Tiny grey feathery foliage, small white or pink blooms. Scent of Southernwood. Slow grower and small plant. B
'Attar of Roses': Soft rose fragrance, dwarf plant with grey/green lobed leaves. Pinkish mauve blooms. B
'Brunswick': Sweet, aromatic foliage. Compact plant with large green leaves. Large flower of light magenta/rose with dark markings. E

'Copthorne': Large leaves, cedar scented. Medium to large mallow, purple blooms, compact but strong plant. B

P. crispum: Many forms including small-leaved and variegated-leaved varieties. Lemon scent, medium mauve flowers. Small-leaved varieties are compact. B

P. dichondraefolium: Difficult to grow but a lovely little plant with small grey foliage, white flowers. Lavender fragrance. E

'Endsleigh': Lobed leaves with dark centre. Pepper scent. Rose blooms with markings. Prostrate habit. E

P. fragile: Small grey heart-shaped leaves, pleasantly aromatic. Medium flowers of pale cream on long stems. Very compact if grown in good light. E

P. x fragrans: Small grey/green leaves, lovely spicy pine scent, some say nutmeg or even eucalyptus. Small compact plant with small white flowers. There are various variegated forms and also crosses; 'Lillian Pottinger' an excellent form. B

'Godfrey's Pride': Large soft yet strong leaves on a vigorous plant, minty/rose scent. Flowers pink. E

P. graveolens: The most popular, strong and vigorous with a rose/citrus scent. Attractive cut leaves, small pale pink flowers with purple markings; there are variegated forms, 'Lady Plymouth' and 'Grey Lady Plymouth'. B

P. grossularioides: A straggly plant with small magenta flowers. Dark leaves and a scent of coconut. E

'Joy Lucille': Rose/peppermint scent. Soft grey, cut leaves. White flowers with dark veining. B

P. limoneum: Small, dark green leaves, plant can become straggly, medium size blooms of dark mauve/pink. Cinnamon/rose scent. B

'Lemon Fancy': Strong scent of lemon. Medium sized plant and leaves. Mauve flowers. B

'Little Gem': Nice dwarf habit. Spicy rose scent. Lots of mauve flowers. B

'Mabel Grey': The strongest scented of the lemon types. Rough tri-lobed leaves, slow grower. Large flowers for a scented of mauve with markings. B

P. x nervosum: Compact plant with small shiny leaves, lime scent. Mauve flowers. E

P. odoratissimum: Compact, with small apple-scented leaves. White flowers. B

'Prince of Orange': Glossy foliage scented with orange. Erect habit with small leaves. Large mauve flowers. B

P. quercifolium: Many different forms are available. Oak-leaf shaped foliage with dark blotch in centre and pungent scent.

'Princess Alexandra'

Rosebud type Zonal Pelargonium Appleblossom Rosebud

Pelargonium echinatum

Geranium robustum

Deacon 'Arlon'

Pelargonium x. schottii

Ivy-Leaf 'Lila Cascade'

'Plenty'

Some forms trail, others are compact. Large flowers of mauve with darker markings, some have serrated petals. Most B
P. radula (syn. *P. radens*): Upright shrubby plant with rasp-like leaves deeply cut. Scented rose/lemon. Pink flowers. B
P. tomentosum: The flannel-leaved *Pelargonium*. Vigorous plant with large soft flannel-like leaves, the plant tends to grow horizontally. Strong peppermint scent. Quite large white flowers. There is a form known as chocolate peppermint having dark brown leaf markings. B
'Toronto': Large lavender flowers on a compact plant with glossy foliage. Ginger scented. E

THE UNIQUES

This group is certain to have come from, at some time, a form of *Pelargonium* species called *P. fulgidum* although since the first introduction of these species hybrids in the nineteenth century the Uniques have almost certainly gained other species in their make-up as well as perhaps some back-crossing from cultivars. *Pelargonium fulgidum* is a plant with small red flowers, a semi-succulent stem and lobed foliage of pale green colour and clothed in very short downy hairs. Uniques themselves are of a shrubby nature with often quite woody stems. The leaves are large with lobes and are scented, some more pleasantly than others. Most will make large shrubby plants and will require a fair space except for the one or two which are of more dwarf dimensions.

It is not definite how the name Unique arose. In fact it is written that the group as we know it today was not actually the group of plants to which the tag Unique was originally intended – or perhaps they came from a plant called 'Old Unique'? Still, the Uniques are well worth growing whatever their shaded past!

Roots, Foliage and Blooms

Roots
Roots are fibrous with a few of these very strong and thick. Most require a good root run so a fairly large pot is better for a mature specimen.

Foliage

More often than not the leaves are deeply divided usually into five lobes and each with teeth or crenations at the edge. Leaf colour is mid-green to light grey/green. They can be shiny or soft and flannelly. In some hybrids the foliage is ruffled. The size of the leaf is from 1.5 in. (4 cm) to the size of a small hand with the width being half the length in most cases.

Blooms

Small to medium flowers of a single type make up the flower head of between four and eight blooms. Some have narrow petals but some broad. The colour range is wide from white to pink and red, nearly purple and magenta. All with blotchings or markings, generally heavier on the top two petals, in varying degrees and some with light or heavy veinings in similar or contrasting shades.

Cultivation

Generally easy to grow if one bears in mind that as a naturally straggly grower they must be kept in check by pruning – they are not really self-branching so cutting back will encourage more young growth. Pinching out a young plant is essential to make a bushy specimen. Because their tendency is to make woody stems without much foliage lower down the plant, pruning and pinching out is very important. Never, however, prune too hard, especially back to hard wood, because this could cause the plant to suffer and perhaps die. If a plant is in this state it would be far wiser to take a few tip cuttings and start again with the newly rooted piece, keeping it short and stocky by pinching it out at the first potting up stage and when-ever necessary after that. Don't forget that the flower buds will be produced at the tips of the growth so the more tips the more flower production. Remember that if flowers are required at a specific time, pinching out and pruning must stop, if only temporarily, three months before the plant must bloom.

Uniques are fairly greedy feeders and a general liquid feed with a little more nitrogen than used for other types of *Pelargonium* can be given – this will discourage woody growth to a degree. Young plants seem to do better, therefore it is advisable to keep Uniques for only a couple of years or so, unless the plant continues to throw new and healthy growth and enough blooms to please.

A good John Innes No. 2 compost with one-eighth added

drainage material, which had better be of a weighty type to help the stability of a large plant, will be suitable to pot the newly rooted cuttings into. Because of the naturally strong upright habit, a larger rather than smaller pot will be more practical. Depotting will be impractical also for fear of the contents becoming top-heavy in too small a pot. Staking could be necessary if it is planned to grow the plant to its full potential, which could be tall enough to reach the roof of the greenhouse or conservatory in a couple of years. If this is the case the problem of not having flowers and foliage at the lower portion of the plant may not matter so much because other plants could stand at the base with the rest rambling up to the heavens – as it nears the light its production of blooms will be naturally improved. Always remove spent blooms to try to encourage a follow on of flowers on all plants.

Uniques rarely flower during the colder, darker months but come into bloom very early in the summer or even late spring. On the whole their flowering period is one good flush but a few more flowers can be encouraged at other times during the summer months. Uniques will fare well out of doors giving height to a collection of mixed *Pelargonium*, their cedar and pine-scented foliage an added pleasure. They must be brought into a frost-free environment in October, by which time the plants will be quite large; cutting back to a shoot will be needed so that the plants may be accommodated in the greenhouse or in the house. The plant material may be saved for cuttings.

Propagation

Seed

As they are closely allied to the Species the propagation can follow along the same lines. Uniques do not seed often so seed sowing to produce more plants will not be worthwhile and of course the seedling would be of unknown parentage and its performance uncertain. To try to hybridise Uniques is a hit and miss affair due mainly to the undetermined ancestry and is not very often done. For maybe the same reason few notable new varieties have been introduced since the early 1900s apart from some sports.

Cuttings

To take cuttings would seem to be the best way to increase the stock of plants. The conventional method of tip cuttings will prove most successful but they may take a while longer to root than other *Pelargonium*. The summer months are preferred and heat will need to be provided during the cooler, darker months. Root cuttings can also be taken from mature plants. Division is not advised and layering will, due to the upright growth and slender stem, be difficult.

Pests, Diseases and Disorders

Pests and diseases are as for the Species but Uniques do become infested with whitefly given the chance – the means of eradication have been already discussed. Some growers place a 'control' plant in their growing area to signify when whitefly is present. A plant of the Unique type would be ideal for detection – if there isn't any whitefly on the Unique it is a fair bet that the other plants in the growing area are clear too.

Exhibiting

Showing Uniques is not too popular, maybe because they usually grow too large to be transported or it could be that they are unfairly put in the same class as either primary hybrids or in the Scented-leaved class. Recently, one of the national shows did provide a class purely for 'Unique types' and it was well supported. The plants are judged and pointed following the Regal *Pelargonium* system. The general aims and condition of a good show plant are to be followed as others.

As Floral Art subjects the foliage is attractive but usually a size suited to arrangements of normal size, whereas some of the smaller blooms may be used for miniature or petite arrangements. Both flowers and leaves will last well in water.

Unique Plant List

'Claret Rock Unique': Leaves deeply lobed and aromatic. Claret red flowers. B
'Madam Nonin': Scented foliage and tyrian rose blooms. B
'Purple Unique': Large purple flowers. B
'Rollinson's Unique': Magenta-purple. B
'Scarlet Pet': Strong scented foliage smaller than type. Bright scarlet flowers. B
'White Unique': Large, striking white flowers with some veining of purple. B

UNIQUE HYBRIDS

Unusual but beautiful additions to the Unique types that are difficult to place definitely in either catagory of their parents, are the Unique Hybrids as they are popularly called.

These derive from crosses of *Pelargonium domesticum* (Regals) with some *P. fulgidum* evident and the dominant parent of Uniques, possibly Scarlet, as the main Unique variety.

Miss Frances Hartsook raised these charming hybrids in America during the early 1960s. They are tolerant of extreme heat and sun and have many mainly small to medium Regal style blooms. The flowers' tones are mostly plain but with unusual veining and darker blotching. The colours are salmon to fluorescent orange, purplish red and scarlet or pink. The foliage is similar to the Regals but a little smaller, some have a slight scent and one has faintly variegated leaves at certain times of the year or growth cycle. In stature they can be classed as shrubby but of a more dwarf habit than the Uniques and so are useful to plant in tubs and for general work outside in a sunny spot.

To exhibit these varieties could present a problem in the normal accepted classification because they are not Primary Hybrids nor Regals; nor are they Uniques, so unless the schedule includes them in the Unique Classification it would be best to enquire of the Show Secretary before making a blunder. In the Unique class a plant in good flowering form and general condition would do well, all things considered, mainly due to the hybrid's compact and stocky habit. The early shows, which are usually held for Regals, would be the best time to exhibit bearing in mind the Unique's early flowering performance.

Hybrid Unique Plant List

'Bolero': Large lobed leaves. Purple-rose blooms with brown markings on upper petals. B
'Carefree': Soft red with purple veining and white throat. B
'Hula': Pink flowered with dark blotches on top two petals. B
'Polka': Toothed, lobed leaves with some slight variegation. Orange-red blooms. B
'Voodoo': Somewhat cordate leaf. Dark red blooms with black marks. B

6

Regal Pelargonium

It is a strange fact that whilst people wish to call most *Pelargonium* by the incorrect name, Regals are known extensively as Regal Pelargoniums!

There are many other names for this group such as 'Martha Washington' or 'Lady Washington' – these names obviously came across the Atlantic; 'Show Pelargoniums' – they are indeed showy; and 'Grandiflorums' – another true description. Over the years other names vaguely putting the plant into a sub-grouping were, 'Large Flowered', 'Show', 'Decorative', 'Fancy' and 'Small Flowered'. All these names are now under the one title of 'The Regal Pelargonium'. In the early 1900s a botanist, L. H. Bailey, grouped them under a botanical umbrella as *Pelargonium x domesticum (Bailey)* and *Pelargonium x domesticum* is the title used by botanists and the expert enthusiast when referring to Regals. The species thought to have been used in the beginnings of the Regal are: *P. cucculatum, P. anglosum, P. grandiflorum, P. fulgidum* and possibly many more.

Regals have been in existence for centuries and in the 1880s serious breeding began in Germany, France and England.

This handsome group is very popular as a house plant and is favoured as a show plant. Unfortunately it does not enjoy the British summer out of doors as a bedding subject. Its main fault – it must be admitted it does have a fault or two – is the fact that the initial flowering season is rather short, May and June but a few short flushes of bloom can be encouraged later. Hybridisers are now working to lengthen this flowering period and recently new and improved varieties have been introduced.

ROOTS, FOLIAGE
AND BLOOMS

Roots

The roots of the Regals are of a fibrous type with some being large, long and thick. Normally covered with a bark-like outer skin, very strong and woody especially in a mature plant.

Foliage

The leaves are usually stiff and a little shiny with some short hairs but some have softer more flannelly foliage with more and longer hairs on the surface of leaves and bud casings. The shape, basically, is a rough triangle, often concave or cupped, folded, lobed and divided and with either deep or shallow serrations. In colour, from mid to dark green with no zone or leaf markings present but there are a few variegated forms available. The plant itself is shrubby, leafy and quite large – specimens of over three feet (1 m) in height and width are not rare. This size can be achieved during the second year of growth.

The stems soon become woody at the base of the plant and in time are covered with a thin brownish bark. Stipules are pointed at the tip, green when young, brown on the older stems.

Blooms

These are the glory of the plant – a case of Regal by name, regal by nature. The majority have very large azalea or petunia-shaped single blooms of five petals and there are a few varieties with double blooms. The petals are usually rounded in shape and perhaps ruffled, fringed or curled. The individual florets can be as large as 2–3 in. across (5–7 cm), and as many as fifteen combine to make up the flower-head. The colour forms are endless, true yellow and blue being the only colours not represented. Mauve is the dominant colour due to the presence of *P. cucculatum* in the ancestry. Soft or vibrant shades of red, orange and purple as well as white are available, the purples often so dark and dense that the effect of black is approached. In many cases more than one colour is evident and some have pencilled edges of a different colour or white. The petals are frequently marked with veining or blotching of different or darker colours, usually more heavy on the upper two petals. Hybridisers have tried to create a perfectly round bloom – there are some where this is almost achieved but most still show the true *Pelargonium* flower formation in the florets.

CULTIVATION

Some find it a little difficult to produce a good Regal, mainly this is due to not being aware that the flowering period is short, the plants soon become woody and temperature for developing flower buds and growing good foliage has to be slightly modified from the normal cultivation of the rest of the genus. Compost should be of a soil-based type with a small amount of added drainage to the John Innes No. 2 formula. Some growers do prefer a peat-based compost for growing the cultivars but it is thought the problems of keeping large, heavy Regals stable on the bench, the shortening of the life expectancy of the plant due to the escalation of growth and the chance of introducing more pests are poor reasons for adopting this type of compost for *Pelargonium*.

By the second season the plant will have achieved proportions that will need a 6 in. pot (15 cm) at least but if the plant is destined for the show bench it will not be possible to exceed this size. Root pruning will keep the root-ball to a more manageable size and should be done in very early spring before flower bud development is in the final stages or at the end of the flowering season when at the same time Regals may be plunged in a cold frame for the remainder of the summer. Pruning of the top growth will be needed during its life – using a clean, sharp knife, trim and neaten the plant to shape not forgetting that pruning is not recommended at the cost of flowers. If the plant has been neglected and is of a lanky shape with hard wood at the base, it may be difficult to prune (secateurs will certainly be required) – to encourage new young shoots in healthy abundance and to take new cuttings could be the best plan, discarding the old plant after the cuttings have rooted.

Regals require a different winter temperature programme because the flower buds form at below 60°F (15°C). Flower-bud formation is more successful if around 53–58°F (11–14°C) and a short-day period is maintained for six to eight weeks. Flowers take about three months to form and bloom. Other plant growth is more pronounced at above 60°F (15°C). It is clear that Regals prefer to be growing gradually during the winter months and, in fact, if good Regals are required this is important but keeping the temperature above 60°F (15°C) for plant growth is perhaps not as essential as the bud-forming temperature. In a normal heated greenhouse these conditions should be possible during the British winter and spring. Watering will have to be continued and only slightly reduced during winter but feeding should be withheld at this time and until the beginning of the bud-formation period when a quarter strength should be given at first, increasing to full strength just before flowering, using a liquid feed with a high potash formula. By this time water will be required in plenty, depending on the temperature of the growing area. An occasional foliar feed will be welcomed in which a tablespoon (15 ml) of magnesium sulphate (Epsom Salts) to 2 gallons (9 litres) is dissolved. At other times a fertiliser of high nitrogen should be given according to the manufacturer's instructions. Always water with warmed water especially during the winter when tap or fresh water will be very cold if not icy.

Regals do not seem to mind whether pots are clay or plastic so plastic would be more practical. Do not overcrowd Regals and give ample ventilation to the growing house. Plenty of light should be given from the late spring onwards and the plant turned regularly as much as twice a week if it is situated on a windowsill. Turn them one-fifth each time – that way no one side of the plant will get preferential treatment. Shading will be necessary in the height of the summer or perhaps during the spring if there is a spell of sunshine, and to shade the plants individually with newspaper will avoid blooms and tender shoots becoming scorched or faded.

Remove all dead and dying flowers and leaves. In the winter it is best to cut off half way down the petiole and keep a watch for the remaining part to wither and dry off before it is removed – it is advised to treat the peduncle similarly to discourage any tearing of the plant tissue which could lead to stem rot and other fungal problems.

If it is wished to use Regals for an outdoor display it would be a sensible plan to sink the plant complete with pot into the soil where a small amount of drainage material has been added to the hole. This will mean that the plants, when they have finished flowering, can be easily lifted and replaced by other, later flowering types of *Pelargonium*. Regals will not like being bedded out for a long time but do like the outdoor life at times when the summer is warm and dry.

PROPAGATION

Seed

The production of seed on Regals is minimal and unless a breeding plan is envisaged it is not worthwhile. The seedlings will not come true even if any set. Seed is not a commercially viable proposition or all the seed houses would be selling Regal seeds. If any come to hand the usual seed sowing methods will be fine.

Cuttings

This is by far the most promising way to increase the stock.

To take stem cuttings, prepare all the equipment and apparatus as detailed earlier, making sure everything is scrupulously clean. Fill trays or pots with the 50:50 mixture of sieved sphagnum moss peat and drainage material, soak and drain well. Choose a healthy, semi-mature plant and write a name label. Using only the semi-mature stems, take the cutting from just above a node with a shoot evident on the remaining stem if possible. Trim the cutting to just below a node and push into the compost carefully. When the batch is done, place in a shaded place for a few days them remove into the light but not in direct sun. Do not cover with propagator lid and only water when needed by sinking the container in a bath of warmed water. The cutting should strike in two to three weeks depending on the time of year – if the cuttings are taken during the winter, bottom heat will be needed with a soil temperature of about 60°F (15°C). The summer and in particular the month of July will be the best time for cuttings. Pot up into individual pots of about 3 in. (7 cm) initially, a loam-based compost such as John Innes No. 2 will be ideal with a small amount of drainage material mixed well in.

Root cuttings are easy from Regals but will take much longer to produce shoots and roots. Choose a root of thicker proportions than the rest and after carefully cutting the root then potting the mother plant back up continue as advised for root cuttings in the Species *Pelargonium* section.

Division is not advised and layering hardly ever possible because of the plant's growing habit.

HYBRIDISING

The hybridising of Regals is rewarding bearing in mind the need for a longer flowering season, but to create more colours would be difficult to achieve due to the vast range available. There are many dedicated to the breeding of new Regals and it is certain that flower size, growth habit away from the naturally straggly stature and shorter flowering stems as well as varying colours and markings are the prime objects of today's hybridisers. Records, as has been said before, are important, so too is the correct naming and registering of new crosses. A line breeding programme, deciding what is to be achieved or improved upon, should be planned at the outset.

Make sure that the bees and insects have not visited the flower and removed the pollen to another receptive stigma. The careful use of greaseproof paper bags to cover the opening bloom will prevent this but inspect to see if the pollen or stigmas from both plants are ready before transferring the pollen with a camel hair brush or cotton wool bud. Covering again for a day will ensure that insects do not interfere with the new crossing. Evidence of a 'take' will show in a few days when the ovary at the rear of the flower begins to swell and the petals fall, and soon the seed formation will be seen. It is now a case of waiting to see if the seed is good. The five seeds will later turn brown and the base of the seed in its capsule begin to curl upwards, and this is the time to take off the seeds by hand before they are lost amongst the rest of the greenhouse inhabitants – they should come away easily and the feathery awn begin to furl as the seed is picked. Sow the seed straight away in a pot or seed tray of seed compost to which has been added an amount of horticultural sand or perlite. Just begin to turn the pointed end of the seed into the compost then let nature take over, do not cover the pot or tray with a lid or similar. The resultant seedlings will have to be potted up, grown on and kept, sometimes for two years or more, without any feeding or human interference so that the true habit, etc may be judged. If the growth and flowering habits are distinct then maybe a new variety is ready to be shown to the rest of the Fancy. If serious hybridising has not been tried before choose the Zonals for the first efforts.

PESTS, DISEASES
AND DISORDERS

This section may be taken from the *Pelargonium* pest section. Regals are by far the most hospitable section of the cultivars towards whitefly, in fact, it is often the reason why people do not grow them – what a pity that prevention and observation and the restrictions of growing other plants were not thought of initially.

EXHIBITING

Showing Regals is the area where many growers discover their dedication, passion, successes and failures and still continue! Regals are undoubtably the showiest of show plants.

The general idea is that a young plant of about one year will at its first flush of bloom be at its prime – as a plant gets older it will lose the vigour and thus its show potential. This is so true of Regals and it should be planned that a show Regal should not be more than two and a half years to three years old for top-class competition.

When a potential show plant has rooted, as it is being potted up, take out, with a pair of sharp pointed scissors or knife, the growing tip. This will encourage the plant to begin a self branching habit as opposed to the upright trend. Some do not pinch or stop until the plantlet has begun to establish itself in the new compost. It is a matter of choice and perhaps experience with some Regals but the plant will cope with having its stop and potting up in one procedure – perhaps having new compost to relish in takes its mind off stopping! Anyway, only the best of strong plants should be kept so if a plant cannot survive this procedure maybe it is as well it should succumb. Stoppings should continue every alternate month or thereabouts. By the beginning of the show year all plants should be in their final sized pots. The last stop should be three months before the show month, when each tip should be removed. The forming of buds too early is infuriating and it is often only the strong willed who can take off these early buds – any that are showing a good size a full two months before the show should come off, leaving any small buds in place. If the florets have opened and are becoming faded two weeks early, carefully remove the offenders with tweezers, take care not to touch partially opened flowers which should be opened fully on the day with others partly open and with a few unopened buds to follow. This is a sign of good timing. The weather can play havoc with these stopping suggestions; times also differ with varieties so it requires, admittedly, a great deal of experience and luck to get it just right.

It is a good idea to visit as many shows as possible before attempting to enter and getting hooked. Take notes on the varieties that are in favour, and the plants' performances.

Staking is permissible as long as it is as unobtrusive as possible and above all, necessary. Thin green split canes with green or brown garden string or raffia are favoured but don't purchase the brown string which is coated in tar to give it weather protection as this can burn the tissue of the plant.

Send for show schedules in good time and check them thoroughly, noting pot sizes, rules, times of entry and staging, etc. Most schedules are issued months before the show to give the exhibitor time to take stock and plan. Small points such as a map of the town, knowledge of parking facilities, the hall's lighting and layout will be of value. Allow plenty of time both before the event and for staging the exhibit. Water plants before setting out, and take a small sprayer or atomiser. Take care with staging, make sure the pot is clean, the top of the compost tidy, that no dead of flagging blooms or foliage or seed heads are left on, or that pests or diseases have not been overlooked. Make sure there is a label with the name of the plant but if the name is not known write 'name unknown' – no points will be deducted for errors in the naming but in close competition the judge may take correct naming into consideration. The judge may be able to name the plant with the 'unknown' label. Stake the plant if required and dress the plant by carefully adjusting leaves, stems and flowers so that they are shown off to the best advantage. This should be done away from the staging spot as much as possible apart from last minute details. When the schedule asks for groups of plants these should be as evenly matched as possible and positioned on the bench in the most appealing presentation possible to catch the judge's eye.

The Joint National Judging Rules of the British Pelargonium and Geranium Society together with the British and European Geranium Society has a points table; for Regals the points are allotted as follows: a maximum of 30 per cent for cultural quality, 45 per cent for flower heads, 15 per cent for foliage and 10 per cent for staging and display. When a set or group of plants or blooms is staged these will be judged individually and a further ten points maximum per plant or bloom for uniformity and overall effect.

Cut blooms of Regals are usually staged in vases provided by the show organisers – depending on what the schedule states these can be either heads or pips (florets) sometimes with a leaf or two and sometimes in groups of two or more. The points for cut blooms are a possible maximum of 60 per cent for form and colour, 30 per cent for cultural quality and 10 per cent for staging and display, and the extra possible maximum of ten points per plant for a group.

REGAL PLANT LIST

'Aztec': Large pale pink blooms with bronze and strawberry pink markings. Compact plant. There is a form with fringed margins to the petals. A good show variety. B

'Grand Slam': Bushy habit. Flowers rose-red with dark markings. There are other 'Slams' including a lavender form. All good for showing. B

'Hazel Cherry': Cherry-red with almost black blotches. Other forms of 'Hazel' varieties are available. Most are good for the show bench. B

'Love Song': A new variegated foliage variety. Flowers are pink with dark crimson featherings on top two petals, lower petals strawberry pink. E

'Morwenna': Near black blooms. B

'Pompeii': Unusual dark mahogany with pencil line of white at petal margins. Compact plant. E

'Prima Vera': Sugar pink with large white throat, trumpet shaped blooms. B

'Susan Pearce': Mauve blooms with reddish to purple blotchings. B

'White Chiffon': Pure white on bushy plant. B, with care

ANGELS OR DWARF AND MINIATURE PELARGONIUMS

Grouped under the heading of *Pelargonium x domesticum* these are types of various origins.

The name 'miniature' certainly is not appropriate and 'dwarf' is not following the dwarf size classification in most cases. The one thing they have in common is the shape of the blooms – the Americans have called them 'Pansy-Faced' which is good as a general name classification. Most do not look like smaller editions of Regals in fact only one variety comes to mind with perfectly reproduced Regal foliage in miniature proportions. It is a little puzzling where it all began. Some are certainly direct crosses from Regals and one a Regal sport. Many are crosses and back-crosses of *Pelargonium* species, perhaps *P. crispum, P. grossularioides* and a little of other small leaved Scented cultivars.

'Angel' is the common name used in Britain and is the most acceptable for this endearing group. The early Angels were possibly a reintroduction of a type known as 'Angeline', which is thought to have derived from a species called *P. dumosum* (one that seems to have disappeared today or is hiding under a different name!) A *Pelargonium* called Angeline was catalogued in the 1820s, so to call them Angels is an obvious and appropriate name.

In this country the 'Father of the Angels' was undoubtedly Mr Arthur Langley Smith who was a schoolteacher, and he began hybridising in the early 1900s. Since his first introduction which was thought to be between 'The Shar' and *P. crispum* he created a dozen or more notable varieties. As he was living in Catford, London he named one of his first breedings 'Catford Belle'. This was and still is his most famous and is grown by all who have a collection of Angels.

All have single blooms of five petals in a neat simple form mostly white to mauve with a pinkish/mauve, but new varieties are being produced thus giving a revival to the section. Most have dark markings to the petals in the way of blotching and veining, more markings being on the top two petals which are larger than the lower. The flowers are attractive even though some are not large, it is the quantity of blooms that is so staggering for such a small plant. The blooms have an uncanny knack of always appearing to hold themselves in all directions on the umbel and facing outwards so that the whole plant seems well clothed in flowers. A definite plus in their favour is the very long flowering period even outdoors in the worst of British summers. Their diminutive stature makes them ideal for windowsills and where space is at a premium.

The foliage is small, 0.5–1 in. on average (1 cm to 2.5 cm), mid to dark green with no zone, and of a similar shape to Regals (minus the cup shape) and *P. crispum* combined. Some do have a slightly crinkled leaf due in part to the *P. crispum* ancestry. There is, at present, one variegated form which was bred in Holland in 1978. In height they rarely exceed 12 in. (30 cm) and are fairly self branching though a little pinching out will keep them bushy. Cutting back lightly after each flowering will encourage more of the small, thin branches that will soon produce more buds. Normally no flowers are produced in the winter months but early spring will see the start of a long flowering spell. Always remove spent flowers unless a breeding programme is envisaged. There is room for hybridists to extend the colour range by 'dipping into' other hybrids in the *x domesticum* range perhaps.

To take cuttings, for the beginner July is the best month – take tip cuttings from new wood because the older soon becomes woody and will not root.

The only real pest will be the dreaded whitefly but not all Angel varieties are popular with this pest.

Exhibiting is becoming more favoured in the last few years. They make excellent show plants and are gaining popularity with show committees in that separate classes for Angels or Miniature Regals are being provided. The normal maximum pot sizes for these plants are 4.5 – 5 in. (11–12.5 cm). The blooms, when cut, may be used in flower arranging but as the stems are not long it is necessary to cut a sprig rather than an individual bloom – even so they have single blooms that last well in water.

General cultivation is the same as for Regals, so too are the points allocation and the techniques for showing.

DWARF REGAL AND ANGEL PLANT LIST

'Beromünster': Dwarf Regal with pale pink blooms and cerise blotch towards outside of petals. B
'Hemmingstone': Pale mauve blooms with upper petals marked purple. B
'Mauve Duet': Silver mauve with dark reddish purple marks on top petals. B
'Moon Maiden': Round flowers of soft lilac, upper petals deeper marked. B
'Rita Scheen': A variegated variety with pale mauve flowers and maroon blotches. E
'Tip Top Duet': Mauve base colour with top two petals deep maroon. B

'Velvet Duet': Dark purplish-maroon. Tends to scramble. B
'Wayward Angel': Pale mauve with upper markings of light
maroon. B

7

Zonal Pelargonium

The largest and most popular group by far are known as 'Zonals' or as Bailey's term, *Pelargonium x hortorum*. This is the group known widely as 'geraniums'.

About 1730 Dillenius saw a plant in an English garden which was thought to be a cross of *Pelargonium inquinans* and *Pelargonium zonale* but it is now supposed that the first cross of this nature was before that time, about three hundred years ago. The idea of these two species being the ancestors of the Zonal cannot be denied but don't expect them all to have a zone. The origins come from the *Ciconium* section and carry the characteristics of that section being partially woody and partially succulent in the stems. Other species have been introduced into the Zonals over the years and now there are many sub-groups containing cultivars (cultivated varieties) of differing shapes, structures and size both in foliage and bloom. These will all be evaluated later.

ROOTS, FOLIAGE AND BLOOMS

Roots

For the size of top growth in the basic Zonal the amount of fibrous root present is surprisingly small. There are usually one, or perhaps more, main roots which, if allowed, will grow to a great length as they would in the wild, searching for nourishment amongst the rocks and fissures. Apart from the main root the rest are quite brittle and can be broken easily especially when young.

Foliage

This is mainly a circle or heart shape but it varies within the range considerably and has, in most cases, shallow lobes with crenate cuts at the margins. Soft hairs normally cover the whole leaf and the size of the leaf can be between 3 – 4 in.

(7 – 10 cm) in the basic form – they are attached to the main stem in either an opposite or alternate mode with short or long petioles attached to the leaf edge at the base. Veins spread outwards from the area of the leaf's petiole and can be very deep and ribbed. In colour the foliage is from palest yellow/green to darkest green, almost black. Variegations and further colourings are available in some forms. The zone, if present, can range from a thin pencil-line to a broad mark usually half or two thirds into the leaf face, narrowing where it ends at the leaf base. An overall horseshoe is suggested and this is why the plant is sometimes referred to as the 'Horseshoe Geranium'. The zone may take the form of a dark blotch of colouring in the centre or towards the base of the leaf. The zone itself, no matter what shape, is actually a red pigment within the leaf which shows brown when the overriding of the green is present and its true colour when the green (chlorophyll) colouring is missing, as would be evident in some of the ornamental foliaged types.

Blooms

The original cultivar's form is single, the five petals arranged in a zygomorphic or irregular pattern, the top two upper petals held apart from the three lower. Over the years this form has become more regular and modern hybridisers have created many with seemingly perfectly round blooms, the petals even overlapping in some.

In shape they can be wedge-shaped to rounded, sometimes with serrations or notches to a greater or lesser degree. A furled formation is not uncommon, neither is slight waving. The overall shape of the individual flower (floret) is normally flat but in some instances it is slightly convex and in some extremely concave. The blooms, having more than five petals and up to eight, as a general guide, are known as semi-double; those with more are double forms. Doubling is when the reproductive organs form into petals – therefore, this type are often sterile. The size of the floret can be from 0.25in. to over 3 in. (6 mm – 70 mm). The individual florets (or pips) are held in an umbrel containing from five to two hundred pips – this whole structure is called the truss or the flower head. Colour is of a vast range, from white, pinks, reds, salmon, magenta, pale lavender to orange. As yet only two cultivars of a yellow shade, which must be classed as cream, and no true blue shades, have been bred. The petals, in certain lights, often

115

appear fluorescent and luminous giving off a brilliant hue, particularly in the evening light. More than one colour is often present, a base colour which will be shown through to the reverse of petals and an overlaid colour, the latter often in stripes, smudges or blotches. Veining and central markings or 'eyes' in the blooms are darker or lighter than the flower tones. Petals of some are enhanced with a thin or bold line at the edges, or speckles, usually becoming more dense toward the centre or eye. All these combinations and effects are present in the single, semi–double or double forms.

CULTIVATION

Zonals must be the easiest of the *Pelargonium* to grow, and perhaps this is why they are the most popular. There are not many summer gardens that do not support the genus. For bedding in the British summer they are fairly tolerant but enjoy a hot dry spell rather than a cool damp one.

Their use for bedding will be improved if the grower becomes more selective in choosing varieties that are particularly suitable – a few are excellent for outside work but some can be difficult if the weather does not suit. A list recommending varieties for outside use is included with the plant list. If the plants are brought to the garden from the greenhouse in the late spring, the sinking of the pot (preferably clay) into the garden soil will be advantageous for three reasons:

1.   The plant will not be damaged on planting and so will adjust to the new environment more quickly.
2.   It may be that the garden soil is not suitable in its nature, perhaps too rich or too acid.
3.   At the time when the plants should be removed to a frost-free place in the early autumn the removal will be quicker, less damaging and the plant may be left for a while until time permits the after-care treatment.

This after-care comprises a general tidy of dead and dying or damaged foliage and blooms, a look for slugs, snails and insects and also signs of disease. The plants will most likely be very tall and lush. It will be better if they are cut back, with a clean sharp knife to at least half. Finish the trimming on each stem just above a leaf node to prevent any die back of the stem and to allow a new growth bud to come from the node. Keep under the bench in the shade for a few days, then water and bring into the light so that the plants may be enjoyed for the winter months. Any good clean healthy plant material suitable for cutting may be struck but a little bottom heat will be required further into the autumn and winter months.

In the home, greenhouse or conservatory Zonals will be grown in pots. Plastic pots are fine although clay types are better for the plant. From the rooted cutting stage they should be potted up into John Innes No. 2 with an amount of horticultural sand or shingle. The proportions will depend on the variety or type of Zonal but six parts compost to one part grit is suitable. When potting on do not use too large a pot – it is better for the plant and offers a way of introducing fresh compost regularly if the plant is repotted into one size larger as it fills each pot. Semi-mature to mature plants can be grown in a large pot but in proportion to the plant size – remember if showing that the usual maximum size for basic Zonals is 6 in. (15 cm). Depotting a potential show plant at the beginning of a show season will be disastrous. This procedure can be undertaken when necessary by removing the plant from its container, pruning the major root and only a few of the truly fibrous ones then placing the plant back in the same pot or one size smaller using a mix of the basic compost. The compost should be tapped around the roots as the pot is filled, watered and drained well and shaded for a few days. This is best done in the early spring just before growth is expected to begin.

Zonals enjoy plenty of warmth but not too bright a sun shining through glass, so some shading will be necessary in the greenhouse. Adequate light must be provided during at least the growing period and as far as possible at all times especially if plants are to bloom in the winter. To encourage winter flowering the temperature must be 48–50°F (10°C). To overwinter the plants by just keeping them going a temperature of 40°F (5°C) will be sufficient. A position in a well-lit, not too warm, spare room in the house will enable non-greenhouse owners to overwinter plants. It will be beneficial and less trouble if they are kept only just growing by not feeding and watering frugally until the late spring when watering should be gradually increased and half-strength liquid feed given weekly. Increase to half-strength feed at each watering when the weather and light improves. Use any good quality proprietary fertiliser containing a higher proportion of nitrogen for the first two weeks to assist in the growth of roots, stem and foliage and change to one with a higher potash content to encourage the formation of flower buds. When all signs of frost are past the plants can be placed outside to enjoy the fresh air and natural watering – it will not be long before their preference will become apparent. There could be a need to 'harden off' (as it is known) the plants before placing them outside. If the plants have been in a warm place it will be necessary to put them outside for a few days, bringing them in each evening, especially from the cold spring nights, until they are accustomed to the temperatures outdoors.

Always remove dead or dying flowers and leaves and damaged parts, even in outside positions. Rotting vegetation of this kind will only encourage pests and diseases.

Watering should be kept to a minimum (warmed water preferred) – tap water will be far superior to dirty rain water, so check the condition of the rain water butt frequently. When using tap water it will be best left overnight so that some of the chemicals can be released.

Never over-water or over-feed the *Pelargonium* – it is a case of, if in doubt, don't. If a mature plant is required for blooming through the winter the best plan is to take a cutting during the autumn and when potted up, remove all the flower buds until the following autumn after which they should be left on – the plant should bloom well during the winter with feeding, watering and temperature as advised above. Some varieties will flower quite easily up until and sometimes through the dark winter months without this special attention, in particular the miniature types – there is no reason why they should not be allowed to do this unless they are next year's show plants when blooming at this time will weaken and spoil the shape of the plant.

Growing a standard *Pelargonium* is a challenge and will not only highlight the summer bedding scheme but reward the effort with creditable comments. It is important to choose a variety that is not particularly self-branching in habit. A newly rooted cutting that has not been stopped and that shows a good, vigorous stem is required. Pot on into a good compost and into a 4 in. or 5 in. pot (10 or 13 cm) at once. Encourage the straight upright growth, staking as becomes necessary and taking off carefully any side shoots that develop. Leave the foliage on the stem, this will serve the plant with the energy-giving means through the leaves. Use stakes all the time, replacing or adjusting as necessary. A word about the tips of stakes – to save accidents it is advised that a small pot or similar be fixed on the top of the stake as it is so easy to bend to tend to the plant and receive a good poke in the eye or face. The ultimate length of the stake or cane should be enough to continue into the head of the plant when the training is completed.

Take up the main stem, checking for uniformity of thickness as well as straightness at all times. When the desired height is reached – this may be any reasonable height but remember the anticipated head size – take out the first stop. The early spring will be the best time to aim for with this stopping or pinching out thus giving the head plenty of time to develop during the growing season. Stem foliage may now be removed. The temptation to allow the standard to bloom the first summer will be strong but it will be worth the wait to postpone attempts to encourage flowering and in fact it will be beneficial to remove buds as they are formed, until the second year. Continue stopping every other month and treat the standard, once it has begun forming the head, as any normal Zonal. Standards may be of quarter, half or full proportions, the size of the head being the guide for showing and the final pot size usually 7 in. (18 cm).

PROPAGATION

Seeds

Seeds will be produced by the plants at the expense of flowers and growth, so unless a breeding programme is planned it is best to remove all forming seed. If seed is to be sown it is best as soon as it is harvested and still with the awn attached. Make up a 50:50 mixture of seed compost with perlite or other drainage material. Place in a pot or tray pressing down lightly with the base of a clean pot or tray, leave in a water bath to soak, remove and drain well. Write a name label, then just push the pointed end of the seed into the compost leaving the coiled feathery awn to drill its way into the compost to the required depth. Place in a warm place or on gentle heat if necessary. The cotyledons will appear from thirty-six hours and may take some months but this is rare in Zonals. When large enough to handle by the cotyledon and when a set of true leaves are growing, pot up into the normal potting compost, water, drain and leave in the shade for a day or two. Do not plant in too large a pot.

Cuttings

Stem cuttings may be taken at any time when the air temperatures are warm or when a heated propagator is at hand. Make up a 50:50 mixture of sieved sphagnum moss peat and perlite, vermiculite or horticultural sand, place in the pot or tray pressing down lightly. Stand in the water bath; drain. Choose a healthy stock plant, one that has had its flowers removed a few weeks previously if possible, and write a label for the batch of cuttings. From just above the node on the plant, preferably with a shoot showing at the node, remove a piece from the stock plant of about 2 – 3 in. (5 – 8 cm) in length with a razor-sharp knife. Trim the cutting to just below a leaf node and cut straight across. Take off all leaves except the top two by pressing in an upward rather than a downward manner to prevent any tearing of the stem tissue – failing this the petiole may be cut in half, leaving the piece to dry and drop away. Remove any stipules that may be below soil level. Press the end of the cutting into the soil holding it as far as possible by the leaves, to about a depth of 1 – 1.5in. (2.5 – 4 cm). Tap gently to settle the compost. Leave in a shaded place for a few days and then bring out into the light but not direct sunlight. If necessary place on a heated propagator, without a cover. The cuttings will begin to root within fourteen days. This root forming time will vary considerably and can take months. Generally the unusual flowered types and the ornamental foliage types with a chlorophyll deficiency will take longer than the basic types.

Layering is possible but as most root well with conventional methods it is not worthwhile.

Although the method of taking tip cuttings is the most successful and easiest it is possible to take cuttings from right down the stem of Zonals, using material that is not too old or near the base of the plant. These are called leaf axil cuttings and may be cut from the plant in three ways as the diagrams illustrate; thereafter they are treated in the same way as tip cuttings. Often the length of rooting time will be longer but they can produce better plants, often branching at soil level. There is a way to take cuttings from the very tip of the growing point (meristem) – it is fiddly to accomplish because it is so small and great care has to be taken regarding hygiene. It is a method used commercially because the risk of disease is lessened with such a small area of plant material used and the process is carried out in clinical conditions. Recently there

Simple tip and stem cuttings of Pelargonium.

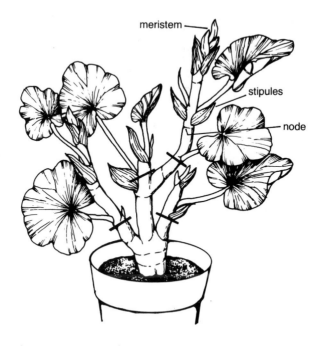

meristem

stipules

node

Leaf axil cuttings

have also been new ways of producing plants, including the *Pelargonium*, by a method known as micro-culture. This is where a minute piece of plant tissue is propagated in a test tube within a sterile situation. Of course this is done commercially also but it is worth a mention because there is on the market a kit and also the rooted plant material for the amateur to try out. The plants produced in this way are virus free and so are perfectly healthy – the worry is that when a plant does come into contact with other sickly plants it seems to have little or no resistance.

HYBRIDISING

The Zonals are the ideal group on which to begin for they offer reliable material and the seed sets readily. It is important that a reason for hybridising is established. What is to be improved – flower colour, leaf colour, shape of both, size of plant, pest and disease resistance? These are some points to be considered. It has been mentioned before that adequate and

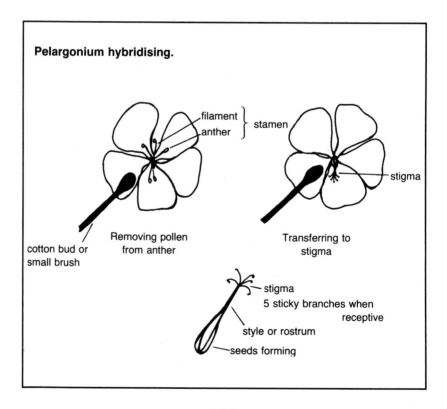

Pelargonium hybridising.

detailed records must be kept. Having chosen the plant that is to act as seed parent inspect carefully the proposed pollen bearer and with a small camel hair brush or a cotton wool bud gather the ripe pollen and transfer it to the receptive stigma which will be sticky and to which the pollen will adhere. After about ten days the receptacle at the back of the stigma will begin to swell and the seed pod similarly. After about thirty days the seed will be ripe and this is the time to watch for their detaching from the base of the rostrum and flicking to the top, remaining there only a short while attached to the tip of the rostrum before falling off – if they are not gathered they will drop to the ground. The seeds should be just firm, and light to dark brown. Sow straight away.

It will be that hundreds or thousands of plants will have to be kept, unaffected by pinching out, feeding, etc. before the chance of that special new cultivar will be worthy of offering to the Fancy. Occasionally a new worthwhile seedling is produced by chance usually due to the intervention of insects and these should not be discarded. These are known as 'chance seedlings'.

If such a gem is produced either by man or nature, it is important that the registering procedures are carried out. For cultivars the International Registration Authority is the Australian Geranium Society and it is advisable also to register with one's own National Nomenclature Committee: in Great Britain it is the Nomenclature Committee of the British Pelargonium and Geranium Society from whom registration forms may be obtained. A few words on the naming of a new cultivar in the form of some do's and don'ts. Firstly and most importantly, choose a name that has not been used before for that type of cultivar. It must not have the word 'variety' in its name but it may have the word 'variegated'. Words relating to other plants must not be used on their own, for example 'Jasmine' but 'Jasmine Beauty' is permissible. Abbreviations and initials are not recommended such as 'Gt. Height' but 'Great Height' is acceptable. The Mr, Ms or foreign equivalents are not allowed but Mrs, is permissible. Do not use excessively long words, nor a name containing the definite article such as 'The Vicar' – use 'Vicar'. Do not choose a name exaggerating the merit of the plant since this may not always be true, such as 'Biggest and Best' or one word that could mean lots of other cultivars like 'Orange' but 'Orange Beauty' is acceptable. The word Cross(es) or Hybrid(s) should not be used but something 'Cross' as a person's name is all right. If a new plant is a sport then the word as a sport should, where possible, show the relationship between the plants, for example 'Lavender Grand Slam' is the sport of 'Grand Slam'. For a name to be valid it must have been published in printed matter which is distributed and dated, such as *Nurseries' Catalogues*.

PESTS, DISEASES AND DISORDERS

The *Pelargonium* is no more prone to pests and other troubles than most plants as will be seen from the next paragraphs – quite the opposite is usual particularly if the ideals of good cultivation, and cleanliness are adhered to. It must be said though, that Zonals do have the longest list of problems – there are some afflicting purely Zonals and not the rest of the family.

Zonals are not smitten with whitefly seriously – greenfly is more of a problem and this with other aphids has been dealt with in preceding chapters. Red spider mite *(Tetranychus telarius)* which will flourish in a hot, dry greenhouse is a problem and as the weather improves and the atmosphere becomes drier the minute, wingless mites may be found feeding mainly on the lower leaf surfaces, and in a large infestation can be seen as small specks with a covering of webbing on leaves and buds. They have built up a resistance to some chemicals so if chemical cures are to be used choose one especially for the control of spider mites. A cool, clean greenhouse will deter these pests as well as, during very dry spells, a damping down with water of the greenhouse floor.

Leaf gall or bacterial fasciation *(Corynrbacterium fascians)* is a type of plant cancer which is seen as gnarled whitish-green shoots occurring at soil level from the base of the plant's stem. It is generally accepted that there is no cure and all plants should be destroyed with no cuttings taken from affected plants. For the inexperienced and the beginner this is good advice.

Many types of viruses can cause many types of problems, particularly on greenhouse plants. It is unlikely that when growing relatively small numbers of plants virus problems will go undetected. The golden rule, as always, is only buy or accept plants that are clean and healthy. Any plant showing signs of unknown markings or deformities, either in the growth of the foliage or generally, should be given a wide berth. Most virus diseases are uncontrollable, so the only answer is to destroy the affected plant and keep a very watchful eye on the rest of the collection.

Black–leg is caused by a fungus which is soil borne called *Thielaviopsis basicola*. It is most common on new cuttings and can usually be seen as a black area from the base of the cutting up. The unaffected part may be used as a new cutting but remember to sterilise the knife before using again. A damp, old compost is often the reason coupled with damage to the outer tissue of the cutting or plant. Always use new, fresh and sterilised compost.

Botrytis has been mentioned before as has leaf-reddening.

Crook-neck and proliferation ('hen and chickens') and also fasciation are other disorders that might be found with mainly Zonals. These are all show faults if left on the plant. Crook-neck is found on the flowering stem at the point just below the flower head and from where the head will normally straighten up on blooming – sometimes this part of the stem will crack and remain crooked. It may be the result of over-feeding and over-watering and often it is apparent at the beginning of the season. These last two factors are often the cause of prolifer-ation which is the formation of another flower head arising out of the first – in rarer cases it can apply to an extra growing tip either at the meristem or from the flower head or truss.

Lastly, by far the most worrying condition in Zonals is rust, *(Puccinia pelargonii zonalis)* which is a universal problem. The rust spores may be spread by air, water, from other plants and on the hands or clothing of humans who have been in contact with the spores. It is seen almost exclusively on foliage and is first found on the undersides of leaves as small circular yellow spots which increase to about 0.5 in. (1 cm) in diameter, then darken to rust-coloured pustules forming a powder-like ring. It is normally rife in damp, cool conditions either in the greenhouse or out of doors. It will find difficulty affecting plants with hard, thick surfaces such as the Ivy-leaved types. The best control is to remove all affected leaves with as little movement as possible and then destroy the leaves totally. Wash hands and instruments used before inspecting nearby plants. Raise the temperature in the greenhouse if possible, reduce watering and increase air circulation. It will be difficult to offer preventive measures to plants outdoors but removing some of the larger foliage so that air may circulate better will assist, together with a regular spraying of the fungicide mix-ture. There is no perfected cure for rust at present.

EXHIBITING

There are classes in most shows, large and small, for the various forms of Zonals. Depending on the size of the show there will be separate classes for each kind, for example singles and doubles, miniatures and dwarfs, ornamental foliage types and perhaps the various separate flower or foliage forms.

The maximum points for judging Zonals other than ornamental foliage kinds, are 30 per cent for cultural quality, 15 per cent for foliage, 45 per cent for blooms and 10 per cent for staging and display.

All the general rules and advice offered previously are applicable here plus the stopping and disbudding dates which are, as a guide, as follows. The alternate monthly stoppings are required for most Zonals for the period of up to three months before the show. Disbudding is advised for up to two months or so before the show, depending on type and variety to some extent. The removal of all open flowers is recommended until two weeks before the date. Staking, training and encouragement both to flower stalk and plant stem is advised continually to give a rounded plant which is evenly clothed in leaf and flower. Remember the finer points of exhibiting, the checking for pests, the cleaning up of the pot and the writing of a clean readable name label. Label cards are often provided by the show organisers for the exhibitor to write on the name of the variety, neatly, and then they should be placed in front of the pots. Label cards are to deter others from removing labels from the pot as well as informing the public of the plant's name.

Florets and cut blooms are pointed as for Regals and the points system of an additional ten maximum points per item in a collective group applies.

Standards should be grown on a straight clear stem the height of which should be greater than the head of foliage, this means the height of the stem from the soil to the first break or branch of the top growth should measure more than the height of the top growth or head. The head should comply with the rules on proportion for a normal plant as well as the points or judging standards to blooms, cultural quality, etc. Zonals are seen extensively in the Floral Art sections. Those blooms from miniatures and dwarfs to the heavy-headed varieties and the ornamental foliage types require the usual attention to conditioning. Their lasting qualities in water are surprising and they will keep fresh for two weeks or more.

Many forms of Zonals are available and will be evaluated in the next section.

ZONAL PLANT LIST

Single-Flowered Varieties

'Ainsdale Beauty': Rose pink with large white eye. B
'Beatrix Little': Compact plant. Vermilion blooms. B
'Christopher Ley': Strong grower. Large, orange-red flowers. Good for showing. B
'Edward Humphris': White. B
'Feuerriesse': Velvet red blooms of a large size. B
'Highfield Perfecta': Coral-salmon on strong plant. A good show variety. There are many 'Highfield' varieties, most are excellent for showing. B
'Loveliness': Pretty pinkish mauve. B
'Mr Wren': Outstandingly attractive but awkward to grow. White base to petals with clear orange streaks. Ungainly plant at times. E
'Stadt Bern': Brightest red, dark foliage. B
'Willingdon Beauty': Rosy-salmon. B

Semi-Doubles and Doubles

'Ashfield Blaze': Semi-double blooms of bright scarlet. B
'A.M. Mayne': Magenta-purple blooms with orange eye. B
'Creamery': Deep cream flowers, shading towards pale yellow. B
'Queen of Denmark': Semi-double flowers of deep salmon. B
'Regina': Appleblossom pink shadings of salmon pink. Good variety for the show bench. B
'Royal Flush': Good purple flowers on compact plant. B
'Santa Maria': Large luminous blooms of salmon. B
'Shocking': Bright pink flowers. B
'White Pearl Necklace': Semi-double, pure white. B

BIRD'S EGG VARIETIES

The late 1800s saw the introduction of the Bird's Egg types, probably heralding from France. The term 'Bird's Egg' refers to the spotting on the petals which is in a darker colour or shade from the base colour. Petals of the florets are single or double and are of a medium to large size. Foliage is small to medium and pale green with hardly a zone evident. There is a form with a golden yellow leaf. The overall plant size is of a medium nature with a tendency to become leggy so pinching out early is needed.

During the 1980s varieties were bred with striped and flecked or spotted petals. These were at first known as Egg Shell varieties but later have been referred to as Bird's Egg types, a definition not suiting at all. These stripes often extend to a major part or at times to the whole petal and it is not uncommon for one petal to be a clear light colour and another to be a clear dark colour with no flecking at all. The base colour is usually off-white with pink to vermilion markings.

Bird's Egg Varieties Plant List

'Baby Bird's Egg': A miniature. Palest pink with bold spotting in cerise-purple. E

'Double Bird's Egg': Pale magenta with plum spotting. Double flowers. B

'Plenty': Double flowered, off white with spotting of rose pink. B

'White Bird's Egg': White blooms with rose spotting. Single. B

CACTUS OR POINSETTIA FLOWERED

It is not certain when these kinds were introduced, maybe around the end of the 1800s and whether it was the English or the French who can claim the 'first' is also disputable.

Known as Cactus-flowered it is easy to see how the name Poinsettia came to be used, mainly in America, to describe the twisted and lengthways furled and quilled petals that may be in single or double formation. On each head there are up to twenty blooms, arranged neatly and for all to see. Some of the petals are medium to long but most are on the small side. Nevertheless this distinct type is worth growing just for the velvety look of the petals. There are reds, corals, pinks and whites among the colours. The foliage is light green with only a faint zone, if any, and there is a variegated form. Stems are long and not sturdy so pinching out is necessary. For the show bench they are not popular which is sad because when included in displays they always invite curious questions from the public. The blooms will cut and last well in water. Many of the present-day varieties were bred in America.

Cactus-Flowered Varieties
Plant List

'Attraction': Camellia rose with striped marks of coral pink. E
'Coronia': Single blooms of pink. B
'Noel': Double, white. B
'Spitfire': Variegated foliage of white and green. Double red flowers. E
'Tangerine': Double vermilion to orange flowers. B

DEACONS

The Deacons or Floribunda *Pelargonium* were introduced in 1970 by the Reverend (as he was then) S.P. Stringer of Suffolk, England. They were derived by selective breeding from a miniature and an Ivy-leaved type. With this parentage one can imagine why most have a compact, bushy habit which makes them popular on the show bench, for outside work in beds, window-boxes, tubs and in the top of hanging baskets if desired. In the greenhouse they can fill a 5 in. (13 cm) pot or larger, still keeping the good shape, and can be equally happy in a smaller-sized pot if necessary. It seems that the larger the pot the larger the Deacon. So popular are they for showing that over the last few years major shows have provided classes exclusively for Deacons. Being easy to transport, to shape and stage their popularity can only continue. Not much in the way of pinching out is required, blooms seem to form in a shorter time from pinching out and flowering usually begins earlier in the season than most Zonals.

Six varieties were first introduced and by the 1980s twenty-four Deacons were catalogued by most specialist nurseries. After this the now Canon Stringer ended the strain and continued on working with miniatures. He died in 1986 after giving the Fancy a multitude of marvellous varieties.

All the colours of the Zonal range are included and all the varieties are of a double form. Many florets make up the spherical flower head and the continuous flowering period of the Deacons is difficult to match. The foliage is medium in size and mainly mid-green and some have zones. There is a variegated Deacon but the markings are not too stable so that at times the faint bi-colour of the leaf reverts.

In shape, neatness and uniformity is the common factor, though some of his later additions are somewhat straggly – these are easily tamed. All are gross feeders so a liquid fertiliser must be given during the growing season and blooming period. Cuttings root readily and may be taken at any time.

As cut flowers they are perfect for the medium to large arrangements, lasting well and not dropping petals.

Deacon Varieties Plant List

'Deacon Arlon': White with greenish centre in bud. E
'Deacon Barbeque': Luminous pink. B
'Deacon Bonanza': Pale magenta. B
'Deacon Finale': Deep burgundy red. B
'Deacon Lilac Mist': Pale lilac with some shading. B
'Deacon Peacock': Variegated foliage. Bright orange flowers. E
'Deacon Picotee': White with some purple staining at petal margins. B
'Deacon Sunburst': Soft orange. B
'Deacon Trousseau': Shaded pink hues. B

FIATS

It is worth mentioning this group if only for the unusual petal shape of some of the Fiats. Frenchman Paul Bruant in about 1870 began a strain of extremely thick-stemmed, coarse-leaved plants which grew strong flower stems and had a vigorous nature without being leggy or too large. Many varieties today owe their strong growth to the 'Bruant' strain but the serrated edge to the petals of some varieties is the interesting variation of the Zonals. The foliage is soft to the touch, mid-green to grey-green and covered generously with short fine silky hairs. No zone, or at least very little, is shown. Shades of corals and oranges dominate the semi-double or double blooms – the florets are quite large and many make up the large, heavy truss.

The plants often listed today as 'Fiats' have one or two varieties that might be credited to other groups such as the

Irenes due to the hybridising patterns many years ago but those with the serrated edges and those with the Bruant tendency are known generally as Fiats.

Fiat Varieties Plant List

'Enchantress Fiat': Soft salmon pink. B
'Fiat': Orange to pink shade. B
'Fiat Princess': Off white with shades of salmon. Serrated petals. B
'Fiat Queen': Coral. Serrated petals. B
'Royal Fiat': Shrimp pink with some shading at edges. Serrated petals. B

IRENES

If it is a large garden, greenhouse or conservatory that cries out for large plants to fill it quickly then the Irenes are the answer – an American strain of plants first introduced in 1942. The originator was Charles Behringer of Ohio who introduced Irene. This cultivar was of uncertain parentage and it was named after Behringer's wife.

It was not until the 1950s that the strain was added to by other breeders out of Irene initially. The prime work was done in this field by Hartsook and Bode during the 1960s. Later Irene types are said to be out of Irene and some Bruant-type crosses. This is very evident with the usually enormous mid-green foliage sometimes of cabbage-like proportions and long lasting strong, rounded heads of many, many florets on stiff flower stalks. The plants themselves possess thick strong stems and are self branching in the main only becoming leggy with insufficient light and incorrect feeding. They are hungry feeders and like more water than the basic Zonals. Flower colour is from white to pink, salmon, orange, scarlet, crimson and bluish-red. Many of the colours are tones of other colours so care should be taken when selecting varieties if a definite mixture of different colours is required. Winter flowering is possible but as the plants are large and will take up a good deal of room the greenhouse space may be needed for the winter storage of the plants from outside. One or two will make a welcome display though, using the method for enticing plants into winter flowering.

The large flower heads are good as cut flowers both for

arrangements and for the showing in the cut bloom class.

Cuttings are easy but the leaves attached to the cutting may have to be cut by half so that over-crowding of foliage in the seed tray does not happen. Pot up into a John Innes No.2 in 3.5 or 4 in. (9 – 10 cm) pots straight away – added drainage is normally needed. Keep them in an airy part of the greenhouse, the large foliage and dense, self branching habit does not lend itself to the passing of air through the plant in an enclosed situation and this can lead to stem rot and other fungal problems.

Irene Varieties Plant List

'Cardinal': Cardinal red. B
'Dark red Irene': Blood red. B
'Electra': Bluish-rose shade. B
'Irene': Medium red. B
'Party Dress': Baby pink. B
'Penny': Mauvy-pink with white centre. B
'Sentinel': White B
'Windsong': Coral shade. B

NOSEGAYS

This section of Zonals is small, in fact, it is one of the collection of types that is known today as one thing but whose title was, a century ago, used for a different race. The early 1800s was when Nosegays were thought to be the popular name for *Geranium crenatum* but twenty years later it was claimed that Nosegay was the popular name for *Geranium fothergillium (Geranium* to be read as today's *Pelargonium)* and maybe there were two lines of descent of the Nosegays of the latter nineteenth century.

There are only six or eight varieties today and almost all known by their colour name with two ornamental leaved ones that can be included. A century ago, it seems, there were more in this group, but alas they must have been lost in time, or perhaps they are now thriving under different names altogether.

The name Nosegay refers to the posy-like heads of blooms held high above the foliage on strong, slender stalks. Each flower is fairly small and comprises five medium to small, slender petals and the individual flowers open at about the same time giving this 'posy' shape to the long lasting heads.

For single-flowered blooms they do not shatter easily and are suitable for flower arranging, the extra long stems being a distinct advantage. Foliage colour is mid-green with no real zone. In the two ornamental leaved varieties the marking is of a vague butterfly pattern.

Generally the Nosegay's habit is upright with a tendency to be leggy – early pinching out will help to form a more bushy specimen. As the leaves are of only a medium size the plants do not look over-covered with foliage – attention to light and feeding will help improve the plant.

Nosegay Varieties Plant List

'A Happy Thought': Green leaves with large butterfly marking in cream. Single red blooms. B
'Crimson Nosegay': B
'Pink Nosegay': B
'Pink Happy Thought': Same as above but with mauve-pink blooms. B
'Salmon Nosegay': B
'Scarlet Nosegay': B
'White Nosegay': B

ROSEBUD OR NOISETTE

Another type of the last century, the Rosebud or Noisette has so many petals arranged in such a tight manner in the bud that it becomes impossible for the flower to open properly and so each floret remains partially closed so giving the appearance of a tiny, half-opened rosebud.

The colour range is basic – red, scarlet, pink, magenta and the 'impulse' seller, Appleblossom Rosebud, having white petals with pink outer edges, resembling perfectly the blossom from which it gets its name.

Foliage is medium-sized and of light green with little or no evidence of a zone. Somewhat tall and straggly in performance, they need careful pinching out at an early stage.

Rosebud Varieties Plant List

'Appleblossom Rosebud': The most popular. White bloom with a greenish centre and pink edging to petals. B
'Pink Rambler': Deep pink. B

'Purple Rambler': Tyrian Purple. B
'Rosebud Supreme': Large trusses of blood red. Good to cut for flower arranging. B
'Wedding Royale': Open rosebud of pale pink. Golden-bronze foliage. B

STELLAR OR STAPH

The original Stellars were raised by Ted Both of Australia and introduced into Europe in 1966. These have star-shaped leaves, sometimes with a zone and there are a couple of ornamental foliage types.

Leaf size is medium and the plant will be well clothed in foliage which, at times, leads to the centre being so dense that undetected fungal problems may strike – careful removal of the inner leaves to open up the plant will help prevent this. Most are of this short jointed habit given good light. The flowers are exquisitely shaped with the two top petals narrow and forked at the edge – the lower petals being wider and wedge shaped with serrated edges. These are the single forms; the doubles have a conglomeration of petals with approximately the top half of the bloom having narrower petals and the lower the broader type.

Showing depends on the treatment given and to an extent the variety. Where possible choose a show schedule offering a class for the Fancy flowered range because most Stellars are not suitable to compete with basic Zonals.

The double flowered forms are useful for some outside work but the quantity of blooms and small trusses restricts their use. However, at least one Stellar type should be included in a collection.

There have been some hybrids in recent years including a race of commercially seed raised varieties called Startel as well as some large growing cultivars from Germany.

FORMOSA HYBRIDS OR
FIVE FINGERED VARIETIES

These are classed next to the Stellars because of the somewhat similar leaf formation although they are not of the same sub-group, being more divided and lobed and looking like long fingers. These hybrids' origin began in Mexico with a plant found growing by Milton H. Arndt of America, in a courtyard of a Mexican hotel. This weird specimen was called *formosum* because it was thought to have found its way via a Japanese ship – later it was realised that the name could not be true and so Holmes C. Miller of America decided 'Five Fingered' was more appropriate.

Only a few have been raised with this unique leaf formation and habit. All have zoned leaves, heavy at times, on an unusually smallish plant of very close nodes so that to keep them free in the centre it is often necessary to remove some of the growth. The petals are mainly semi-double or double and some single but the massed idea of double does not occur always with this type. Some flowers have their petals arranged symmetrically in a daisy fashion and not overlapping at all even though they may possess eight or more petals. The colours are reds, salmon, coral with some veining or tinting. All are slow growing, stocky and miniature.

Stellar Varieties Plant List

'Stellar Arctic Star': Single, large white blooms. B
'Stellar Grenadier': Double crimson. B
'Stellar Hanaford Star': Dark salmon, single. B
'Stellar Snowflake': Double white which tends to 'pink' in summer. B

Fingered Type Plant List

'Formosum': Medium salmon. E
'Red Witch': Double Red. E
'Urchin': A miniature with bright red blooms. E
'Playmate': A miniature, salmon with white tip to petals. E

TULIP-FLOWERED

The most recent introduction to the Zonal forms. Although the first Tulip-flowered was introduced in America in 1966 it was nearly twenty years before it became popular in Britain. Robert and Ralph Andrea raised the first Tulip variety from crosses and back-crosses using a Fiat variety in part of the make-up. The blooms are just semi-double, and the large cup-shaped petals do not open fully but stay in a curved loose bud resembling a Tulip. In colour the first was a carmine salmon shade with paler on the reverse of the petals which, of course, is the side of the petals in view. Flower heads are about 5 in. (12 cm) across and are held on long strong stems – the truss can contain as many as sixty florets of a rather stiff habit so giving the type a good performance rating outdoors as well as in the greenhouse. The petals are almost shatter-proof. The new varieties available are of similar colour tones and do have a tendency to sport one or more normal semi-double formed florets within the flower head. This is obviously something to watch for when choosing a variety for the show bench. The foliage is stiff and shiny with the Zonal-shaped leaf but quite crumpled and lobed and dark green in colour.

The short stems on the fairly vigorous plants give a stocky habit to the plant which will bloom early in the British season.

Tulip-Flowered Varieties Plant List

'Patricia Andrea': Salmon flowers. E
'Pink Pandora': Pink blooms. E
'Red Pandora': Flowers of soft red. E

MISCELLANEOUS TYPES WITH FANCY OR PICOTEE FLOWERS OR UNUSUAL FOLIAGE

As with all things there are some that do not fit in to basic categories or are too few to be allotted a section.

Phlox-flowered are flowers with usually pale petals with a band of darker, but still pale, tone at the base of each petal so that when making up a single round bloom it appears as a halo in the centre of the flower. 'Phlox New Life' is a sport of the ver-milion and white striped variety 'Single New Life' which, if it can be obtained, is a must for all collectors – there is also a very weird double form of 'New Life' but it is very difficult to grow and unless a good clone is discovered does not grow or bloom in a stable manner so is not worth the effort for the beginner.

A race of hybrids from a species which shows the parentage in at least the growth and foliage are the *frutetorum* Hybrids whose pretty coral blooms, either single or double, are held above the foliage on very long, slender flower stems – there is a white form too. Foliage is typical of the species in question – heart-shaped generally – with a dark centre to the dark green glossy leaves. All are fast growers and most have a sprawling habit which makes them a nice change in hanging baskets, or on balconies. One variety has chocolate brown leaves when grown in good light with suffused yellows and light greens as markings.

A variety called 'Kewense' was discovered at Kew Gardens in London and first described in 1930. It has narrow petals of currant red in a single form but of a species formation. Other colours are now possible to obtain, white, pink and scarlet. The leaves are small to medium with a dark zone on the five-lobed leaf.

Miscellaneous and Unclassifiable Types Plant List

'Distinction' (syn. 'One in a Ring'): Small dark red flowers, single. A narrow zone very near the edge of the light green wavy leaf. B

'Floral Cascade': Double flowered *frutetorum* with coral flowers and dark foliage. B

'Kewense': Forms of pink, scarlet, red. Dark leaves. B

'Magic Lantern': Single *frutetorum* with coral flowers and multi-toned leaves. B

'Mauretania': Phlox type. Whitish blooms with a salmon halo. B

'Phlox New Life': Small single white to blush pink, Phlox type pinkish coral halo. B

'Single New Life': Very old variety from around 1869, thought to have been a sport from 'Vesuvius'. Definite stripes of vermilion and off white sometimes single colour on alternate petals or half stripes and half plain colour on same petal, no two petals are marked the same. E

'Skelly's Pride': Hybrid with very shiny, dark green leaves and serrated or fringed petals of salmon, single. B

'The Prostrate Boar': A *frutetorum* type, smallish dark leaves and a scrambling habit. The coral, single flowers are held above the plant on long stems giving an appearance of hosts of moths above the plant. B

F1 HYBRIDS

If any of the many types of *Pelargonium* have ever been accepted with mixed feeling it must be the F1 Hybrids. What is an F1 Hybrid? It is a plant produced in the first generation of breeding from a cross between two selected parents of true breeding cultivars. The seed from an F1 will not come true in any further breedings. The main advantages are that the seed will germinate quicker and more evenly and because a certain amount of disease resistance is incorporated in the new seedling, fewer losses will be experienced; plants will grow more quickly and evenly and so will attain maturity at the same time; they will save maintaining and keeping stocks for cuttings and the flowers will be larger and more in quantity. In a nutshell time, therefore, money will be saved, heat in the greenhouse will not be needed until January in the northern hemisphere, disease will be kept to a minimum and the result will be more uniform and just what the public ordered.

Unfortunately these so-called assets are not what everyone wants. Some like to keep plants the year through, they enjoy their plants flowering at different times, they don't want all the plants to look the same as two peas in a pod and with the F1 Pelargonium Hybrid, variety is very limited. There are no Regals, Angels, true miniatures, few Fancy-flowered, no Ornamental Foliage, hardly a Double or an Ivy-leaved and the F1's do have this infuriating habit of forming seed heads so that the plants can look like 'sputniks'. With these *Pelargonium* the boast that plants flower earlier and better is not always true – it is found to be difficult to have blooms on a young plant inside three months whereas this is possible with cultivars.

Before the seed firms who have spent millions on this breeding exercise begin to shout back let's set the balance. There is definitely a place for the F1 in today's garden and greenhouse – this is evident from the hundreds of customers at garden centres grasping their little plantlets by the dozen. Many people do not have time or the know-how to produce plants from cuttings and they should not be denied the pleasure of seeing a plant reach maturity and bloom with hardly any effort on their part. As long as they grow *Pelargonium* the Fancy should be happy. It may guide these folk towards keeping the cultivars one day and that is what it is all about. The F1's are usually grown in the public parks and gardens by the thousand – they all have to grow at the same rate and flower at the same time to save man-hours and give a first-class result. If they are grown as a bedding plant and treated like most bedding subjects, as annuals, F1 hybrids do have a place in our 'world'. It is such a pity that seed houses insist on calling them 'geraniums' when it should be *Pelargonium* and that those responsible for naming these new varieties do not consult the Offices of Registration to see if there are any long standing cultivars with the same name as the one proposed for the new F1.

There are also F2 Hybrids often called 'open pollinated varieties' – these are not so expensive to buy as the F1's but do not give such a guaranteed result.

The history of the F1 Hybrid began in the States in the 1950s with an expensive and intensive research programme which, in 1967, was fulfilled with the introduction by Pan-American Seed of the first F1 called 'Carefree'. 'Carefree' then was joined by separate colours and was known as the 'Carefree Strain'. Since then many new strains have been produced to cater for the increasing market.

To grow from F1 or F2 seed sow from January until the end of February to achieve flowering plants in the coming summer. Sow in a pot or tray of good quality seed compost and germinate at a temperature of 72–75°F (23–25°C). If wished the seeds may be planted in a warm dark place such as an airing cupboard, until the seed shows the first signs of germination, a constant watch must be kept for this. Some advise sowing on to damp tissue or paper-towelling – this is fine as long as a watch is kept for the first seed to germinate, then the shooting seed will have to be transferred, very carefully, to a seed compost. When they have grown cotyledons and then their first pair of true leaves, carefully, whilst handling them by the leaves, prick out into trays or pots filled with seed compost. Later pot up into pots with John Innes No. 2 compost. Keep them through these stages in a greenhouse or similar environment at a constant temperature. When all danger of frost is past the plants may be planted outside in their permanent positions. F1's are better raised from seed each year, to keep plants over the winter will not only defeat the object of growing F1 Hybrids but they will, no doubt, become leggy. This legginess is common in some varieties and is brought under control by the use of a dwarfing agent which is sprayed onto the plants at specific times during their early growing period. This keeps the plants under a strict uniform growth pattern, encouraging branching and flowering. This chemical is not easy to obtain by the amateur and perhaps this is not a bad thing.

F1 Hybrid Plant List

These will mostly be listed in the series names, the colour name is self-explanatory.

'Breakaway Red': A new type, breaking and branching readily, suitable for baskets, etc. B
'Breakaway Salmon': B
'Pulsar Red': Good strong foliage with zone. Blooms well and has a white eye. B
'Pulsar Rose': B
'Pulsar Salmon': B
'Ringo Deep Scarlet': Good zone and compact variety. B
'Ringo Dolly': An orange and white bi-colour. B
'Ringo Pink': B
'Ringo Rouge': Soft red. B
'Ringo Scarlet': B
'Ringo White': B
'Summer Showers': A new Ivy-leaved type in mixed colours. B

F2 Open Pollinated Plant List

'Bambi Mixed': Early and dwarf. B
'Florist's Strain': Mixture for an economical collection suitable for garden and bedding schemes. B

ORNAMENTAL-FOLIAGED PELARGONIUMS

These are sometimes known as Fancy-leaved or Variegated varieties – neither of these descriptions are accurate so Ornamental-foliaged varieties is a good all-round definition of these varieties with leaf colour other than the basic green with a zone or green without a zone. The two most popular and most colourful are known as Tri-colour and Bi-colour. The description of a tri-colour is a plant with a leaf zone overlaying two or more of the other distinct leaf colours. A bi-colour should have two distinct colours other than the zone, when present, i.e. plants with green and cream or green and white leaves – any zone should only overlap the green part of the leaf.

All these varieties provide wonderful foliage colour so that even without flowers the plant still has attraction. In some bedding schemes the blooms are removed so that the foliage receives all the light and nourishment and grows with even better colour and in carpet bedding the blooms do not detract from the foliage.

There are some lovely golden foliaged or gold and green foliaged varieties. At times and in certain conditions these do lose some of their colour and although for show purposes they are at present listed with the basic Zonals for convenience they will be included as Ornamental-foliaged types. These very light green, sometimes yellow-leaved varieties are in fashion and many new varieties are being introduced with large attractive blooms which may be single or double.

Markings on the Ornamentals can either be a simple blotch with usually a slight or definite division in the centre of the colour, known as 'butterfly marking' or the zone can be so large it is a complete blotch in the centre of the leaf. There are almost black leaves with dark zones or blotches too. The tri and bi-colours are divided into silver or golden depending on whether it is white or cream in the leaf. The bronze forms may or may not have a zone as a separate marking. Some have yellow leaves with a central zone of green – even the stems of some varieties are coloured. There is indeed an abundance of foliage colour available and many just grow Ornamental-foliaged types in a collection.

When purchasing plants it will be evident that even on the same plant the leaf colour will perhaps show a marked difference. Look for a plant with plenty of good colour in the young and semi-mature foliage – don't buy a plant that has any green shoots. This is called 'reverting' and means that the plant has decided to produce a shoot of the same variety as it was before it sported the ornamental shoot. Because it is stronger the green shoot will, in time, take over the rest of the plant and soon the whole plant could be green. Most of the bi-colours and tri-colours originated as a sport shoot from a normal green-leaved plant which was then taken from the mother plant as a cutting, rooted and, if it did not revert readily, was brought into commerce for the enjoyment of others. This is found from time to time and the shoot should be removed and struck or the rest of the plant cut away and the sport left to grow unchecked, removing any green shoots. After the sport has made good growth of the same colouration as the original sport, take cuttings in the usual way. Today some work is being done regarding exposing plants to radiation and it has been found that plants will send out shoots with unusual coloured foliage. This is a fairly new idea and certainly not for the amateur but the future may hold some pleasant surprises in this field. All this modern scientific work is a far cry from the work of Peter Grieve of Culford Hall, Bury St Edmunds, who, in the mid–1850s, gave us many beautiful Ornamental-foliaged varieties that have since the Victorian Era been firm favourites with greenhouse and conservatory owners, not to mention today's gardener.

Many new varieties have been raised and introduced in recent years – some in the miniature and dwarf range, one in the Cactus section, quite a few in the Ivy-leaved, and one or two in the Regals. The golden-leaved varieties also are available in many guises.

For cultivation it must be realised that the growth is not so vigorous as with the basic Zonals. Some may find them difficult to grow and with the lack of chlorophyll in the leaves of the bi-colours and tri-colours a little extra attention to light and feeding and watering is necessary. Good light will improve the leaf colour but strong sunlight will scorch or fade. Place them in a well-lit part of the greenhouse but out of midday sun. Putting the plants outside after the frosts will be less trouble for shading and so on. If any totally white shoots appear these should be removed – they are called 'ghost shoots', have no chlorophyll and will shrivel and die in time, so why let them waste the plant's energy?

Pinching out at the first potting stage will be necessary and also later on for the older varieties – the modern ones do seem to be of a more branching habit as a rule. Taking cuttings is no more difficult than with the basic types but the length of time taking root will be a little longer for most. Pot sizes are really up to the individual but take into account the show maximum size of 6 in. (15 cm) if applicable. Compost can be a John Innes No. 2 and some enthusiasts try using other minerals to produce a more distinct foliage colour, but care should be taken when embarking on this idea even though experimentation is worthwhile. The use of magnesium sulphate (Epsom Salts) and the use of an African Violet (Saintpaulia) fertiliser will enhance some ornamental leaves but do try other feeds and horticultural additives in small doses, keeping a record of the effects. As with all pelargoniums, don't over-water – Ornamentals do not grow as quickly as most Zonals so they do not use up so much nourishment.

On the whole they do not have such large blooms – these may be single, semi-double and double. The Zonal range of colour applies but there are very few white forms.

Exhibiting

The general remarks about exhibiting Zonals apply but the allocation of the points is different in that for the Ornamental-leaved Zonals the maximum of 45 per cent is for the foliage and 15 per cent for the flowers with the staging and display having a maximum of 10 per cent and the cultural quality 30 per cent. From this will be deduced that the foliage is all important on the Ornamental-leaved types and it is possible to do well on the show bench with very little bloom evident. Some shows organise a class for the leaves to be placed on a board, which is normally black – usually six ornamental leaves are requested. These should be of typical colour and size of the variety, clean and bright in colour and named on cards usually supplied.

Ornamental-foliaged pelargoniums are well worth the small amount of extra effort needed to produce a dazzling pot plant with so many forms of leaf patterns and colours.

Ornamental-Foliaged Varieties
Plant List

Tri-Coloured Types
'Dolly Varden': Single, scarlet blooms. Soft leaves of good tri-colour form. B
'Falklands Hero': Pale yellow leaves with good heavy zone in overlay. Single, large red blooms. E
'Henry Cox': Sometimes seen as Mrs or Mr. Small, single, salmon pink blooms. Dark zone in the formation. E
'Mrs Strang': Double orange flowers. B

Bi-Coloured Types
'Chelsea Gem': Green and white foliage. Phlox pink double flowers. B
'Flower of Spring': Green and white leaves. Single scarlet flowers. B
'Frank Headley': Smallish foliage of white and green. Single salmon blooms. B
'Leamington': Cream and white foliage. Single pale cerise flowers. B
'Princess Alexandra': Grey leaves with whitish borders. Double rose-pink flowers. B

Golden or Green-Gold Types

'Bridesmaid': Golden foliage, double salmon pink flowers. E

'Fenton Farm': Golden foliage and light purple single flowers. B

'Morval': Double soft blush pink. Light coloured foliage with bronze zone. Dwarf grower, good show plant. B

'Pink Golden Harry Heiover': Bronze, green leaf. Single pink flowers. B

'Tuesday's Child': Single pink flowers. Leaves yellow-green with bronze zone. B

MINIATURE AND DWARF PELARGONIUMS

During and just after the last war the *Pelargonium* was almost forgotten since most greenhouses were being used for food crops or falling into disrepair because of difficulties in finding building materials, obtaining extra fuel and keeping staff – the greenhouse itself became at risk. However, after the 'dig for Victory' era many people needed a change and turned to growing Ornamentals. At the same time, fortunately for both the greenhouse and the *Pelargonium*, development in the use of aluminium and also the start of the 'do-it-yourself' era brought the dreams of owning a greenhouse nearer for the ordinary person. To erect a greenhouse in one's own back-garden was something of a scoop. What happened next was that greenhouse gardening took off with a bang! Everyone with a greenhouse wanted to grow everything. The 'mini' craze developed as a result of the fact that greenhouses on the market to suit most pockets were a little too small to accomplish this and also that post-war houses and gardens were generally smaller, as were most things then. The *Pelargonium* were no exception. For years we had had plants with large flower heads, leaves and growing habits which were just too big for the small greenhouse, windowsill or balcony to cope with in any quantity, so a few raisers began to develop the dwarf varieties that were already available in small numbers.

From the late 1950s hundreds of miniature and dwarf Zonals came on the scene. Surprisingly the line producing these is over one hundred years old. 'Red Black Vesuvius' and 'Salmon Black Vesuvius' (Red Black being a sport of Vesuvius) are credited with being the first of this type. These two varieties also possess near black leaves so it is easy to see why so many miniatures and dwarfs have this dark foliage.

The height of dwarfs is reckoned to be, as a guide, from soil to top of foliage, not exceeding 8 in. (20 cm) and above 5 in. (13 cm) and for showing should not be in a pot larger than 4.5 in. (11 cm) in diameter. The miniature should not exceed, from the soil to the top of the foliage, 5 in. (13 cm) and for showing the pot should not exceed 3.5 in. (9 cm) in diameter. There are even smaller types called micro-miniatures, which are of the minority and like miniatures and dwarf should have their foliage, flowers and size in proportion and all these in proportion to the pot for showing.

The miniatures and dwarfs all contain the various types as in the Zonal classification, and there are also a few who comply to these sizes in the Ivy-leaved section. Single, semi-double and double flowers are of the same colours as Zonals. Although some have tiny blooms there are some who possess blooms larger in comparison to the basic Zonals.

The care of these types is only marginally more difficult. Diligence with watering, air circulation, winter warmth and routine inspections to remove dead or dying material and detect pests and problems will repay generously with young-looking, healthy plants that will usually bloom throughout the winter months in a light place with an air temperature of 55°F (13°C).

Propagation does need a little more care. Remember that the plant may be only a few inches high and so short-noded that it may be difficult to find any cutting material. Mix up 40 per cent seed compost, perlite and 20 per cent horticultural sand, fill a pot, press down lightly, soak in a water bath and drain well. Choose a healthy plant and write the label. Cut a shoot from the main plant and if it is during the colder months, dust the wound with a fungicide powder. Trim the cutting to just below a node, then cut straight across. Remove leaves and stipules as before and press the cutting into the compost. Do not cover. In the summer they may start to root in two to three weeks; in the colder months bottom heat will be required. Beginners are advised to take cuttings in the summer until some expertise is gained. When the cuttings have rooted put into 2 in. (5 cm) pots and later into the standard size. The potting compost should be a loam-based of John Innes No. 2 formula with one-quarter of drainage material added. Some varieties will not need so much drainage material. Water regularly but only when individual plants require it – do not over-water. Feeding should be minimal, only a quarter strength at each watering, with an occasional change, after the plant has been in the compost for six weeks or so.

Miniatures and dwarfs are not afflicted by any extra pests apart from being a little more prone to black-leg and other fungal troubles. Care with cleanliness and the use of fresh compost and adequate air movement will help to keep these problems at a low level.

Hybridising is still popular and it is becoming increasingly difficult to develop a new plant with new merits. There is a need for new varieties but they must be an improvement on those in cultivation at present.

The exhibiting remarks and the points system is the same as for Zonals.

Miniature Varieties Plant List

'Anna': Single, large flowers of mauve with white eye. B
'Chi-Chi': Single, salmon with darker at centre then a pale eye. B
'Delta': Double neon-pink, narrow petals, slightly furled. B
'Frills': Unusual shaggy blooms of coral. B
'Jayne Eyre': Double dark lavender blooms, dark foliage. Good show plant. B
'Royal Norfolk': Rosy-purple, double. B
'Silver Kewense': Green and white variegated form of Kewense. Red, single flowers. E
'Variegated Petit Pierre' (syn. 'Variegated Kleine Liebling'): Small rose pink blooms and ruffled green and white leaves. E

Dwarf Varieties Plant List

'Billie Read': Double, carmine and magenta flowers. B
'Fantasie': Good white double. B
'Orangeade': Double, brilliant orange booms. B
'Tom Portas': Empire rose, double. B
'Terrence Read': Single, wine-red. B
'Wendy Read': Double pink shading to white. B

8

Ivy-leaved Pelargonium

The Ivy-leaved *(Pelargonium peltatum)*, trailing or basket 'geraniums' as they are popularly named are almost as widely used by the general public as Zonals. Deriving from *P. peltatum (peltate* meaning shield–like due to the petiole being attached, in most varieties, at the centre of the undersurface of the leaf) they are recognisable as being called 'Ivy-leaved' because of their five angular lobed leaves assuming the appearance of Ivy *(Hedera helix)* and even in some varieties smelling vaguely similar. Their natural habit is to scramble either along the ground or up through shrubs so the stems are long, more than 6 ft sometimes (2 m) and long, thin, and infrequently branched. They have been in cultivation for about two hundred and ninety years.

They are used today in nearly every garden as ideal specimens for hanging baskets, window-boxes, tubs, over walls and as pot plants or for a limited type of bedding as summer ground cover, on the flat, or clothing banks; they can also be happy in a scree situation until the frosts begin. They will not climb of their own accord but will, with some means of attachment, be able to climb up a system of trellis or canes and may reach up to the roof of a greenhouse or conservatory to reach 12 ft or more (4 m) if allowed to grow throughout the year.

Crosses from Zonals and Ivy-leaved are not uncommon and there have been some lovely plants with this make-up, most with double flowers and coming from the Continent in the majority of cases. Recently a group called the 'Harlequins' were developed by grafting methods between a red and white striped variety called 'Rouletta' (Mexicarin) and various other varieties of Ivy-leaved. These 'Harlequin' types became popular during the end of the 1970s and the beginning of the 1980s and initially six varieties were introduced.

ROOTS, FOLIAGE AND BLOOMS

Roots

These are of the fibrous kind, fairly coarse and wiry.

Foliage

The leaves are shaped like common ivy and are mid to dark green in colour, thick, glossy or felty, and have small hairs on them – often small hairs are seen on the flower bud casings of the glossy leaved types and some have an aroma when bruised. A zone, often small, is present in some near the centre of the leaf. The stems are long between the nodes and in most varieties, pinching out is needed to encourage them to branch. The older stems are covered with a brown, bark-like skin. Modern varieties have been produced with white stems and these seem to be more floriferous. There are some variegated forms and some with the veins showing as a white netting which is caused by a harmless virus.

Blooms

Flowers may be single or semi–double. The double forms are further classed 'Rosette' types due to the small petals being laid in the bud flat, so that on opening they form a rosette of petals like a small rose. The flower heads are held on long slender stems at most leaf joints. In colour mauve is prominent then pinks and reds with only a few corals and whites. Some markings and veinings can be found and also picotee edging and stripes. About eight blooms are in the truss as a rule. The flowering season is from early summer until autumn – it is not easy to persuade them to bloom through the winter but those with coloured foliage look attractive with their trailing habit as winter specimens.

CULTIVATION

Few are difficult to grow using general *Pelargonium* cultivation procedures. They will need pruning or pinching back hard – the sooner this is done in the young plant the more branches from the base will grow and so produce blooms.

Plenty of light should be available but if any are reaching to the top of the glass structure some damage may occur from bright sunlight. On a windowsill the plant must be turned frequently. Ivies will favour a half-pot, one that is half as tall or nearly so as normal. The trailing tendency could be a problem if the pot is not hung up or stood on another, up-turned pot.

In hanging baskets they will show themselves off and it is up to the individual whether more than one variety is planted into the same basket. It is more attractive to plant the same variety in the same container with three plants round the edge and one in the centre. If showing, unless the schedule states otherwise, it will only be allowed to have one variety in a basket.

When the new cutting is planted a loam-based compost is preferred but some do use a peat-based compost for basket work. This is fine because the basket has to be as light in weight as possible and retain as much moisture too – a basket drains continually so there should be no problem.

Pot size will have to be large enough to accommodate the weight of the sprawling plant but for show it is normally 6 in. (15 cm).

Ivies do like a little more water and also a little more liquid feed every day. A good general fertiliser can be given with a change now and again. To encourage more blooms a feed higher in potash will help.

PROPAGATION

The normal way of propagating from cuttings should be used although due to the straggly growth the taking of good short-noded cuttings may be difficult. Another way is possible with Ivies – this is to break the stem at the node where there should be evidence of a new shoot. This system will allow the stem to break across the node with no need to trim off anything unless the leaves are large, in which case cut them through by half. It is possible, with new shoots showing at each node, to obtain a cutting from each leaf node plus the portion of stem. For the beginner, Ivies are the best type to practise this 'breaking' technique but for the more experienced or more adventurous it is possible with other types of *Pelargonium* too. It must be stressed that experience alone will tell if the plant or type will break easily – if any bending occurs then the plant is not suitable, don't ruin plants by using this method until expertise is gained. The cuttings, when taken, are then placed in the usual cutting compost and treated as for other types.

Layering is possible and may be used on those growing out of doors in the beds.

HYBRIDISING

Ivies have not been raised in very great quantities; many set seed but do seem to throw the parentage characteristics rather than anything very different. The usual procedures are suitable for the hybridising of Ivy-leaved types too.

EXHIBITING

Not so popular as the other types mainly due to the problems of transportation. A good Ivy-leaved on the show bench is a credit to the exhibitor. Always take extra pots on which to stand the plant – this is permissible but they must be scrupulously clean and a good idea is to take brand new pots for this purpose. Staking can be used to train the plant and again this method is accepted except on the odd times when the schedule does not allow. The staking should have been carried out during the months of growing the plant – it is no use trying to arrange a plant, especially the brittle stemmed ivy-leaved, up or over some canes or trellis on the show day!

All normal practices should be adhered to for pot grown Ivy-leaved plants. Hanging baskets and hanging pots are not too popular – again this is due to the difficulty in transporting. The size for hanging baskets is usually a maximum of 12 in. (30 cm) and for hanging pots 6 in. (15 cm), all sizes diameter. Remember that hanging containers should be looked at from eye-level so not only should the plant cover the sides but also the plants should be a good shape on top. Plant up the baskets for the following year in the previous December with semi-mature specimens that have been pinched out – their stopping programme should continue as with normal plants. Aim with the stoppings to creat a complete ball of foliage and flower. The use of sphagnum moss for lining, although advised for home baskets, is not used for exhibiting – it can be messy, drip water and be host to insects and slugs which may escape and find homes on other exhibits! Plastic of an aesthetic colour is suitable.

Transporting has been mentioned as a problem. A base board fitted into the vehicle onto which a pot, large enough to allow the basket to sit in it without being damaged, is screwed, is helpful. This method is also useful to transport other show plants. Others prefer the polystyrene trays complete with holes that are greengrocers' cast-offs. Many showmen are inventors and there have been some ingenious devices used to assist in the safe travel of plants to and from shows.

PESTS, DISEASES AND DISORDERS

Ivy-leaved are very free from problems. Greenfly which will attack the young tips are the worst of the pest and have been dealt with elsewhere. A problem which is a disorder and not a disease is oedema or dropsy. They are small shallow corky marks on the foliage and sometimes on the stems, caused by the swelling of the plant's cells due to the plant taking up more moisture than it can cope with. Often in the early spring, when perhaps the plant has been pruned and more water is being given due to the season, this will occur and show small scar markings – there is no need to worry. The best thing is to take off the worst of the affected leaves – the plant will right its balance soon but, if a little less water is given and no fertiliser at all for a week or two, all will soon be well.

IVY-LEAVED PLANT LIST

'Amethyst': Semi-double blooms of light purple. B
'Beauty of Eastbourne': Double, cerise red. B
'Crocodile': The leaf veins are yellow giving the effect of netting. Single, magenta pink blooms. B
'Galilee': Rose-pink double. B
'Harlequins': Many forms available. They have in common, striped petals with white and another colour or shade. B
'La France': Semi-double mauve with purple featherings. B
'L'Elegante': Variegated foliage shading pink in dry, light conditions. Single, off-white flowers with purple veinings. B
'Magaluf': Cream and green variegated foliage. Neon-rose petals with crimson marks on upper two. B

'Rio Grande': Deepest maroon, almost black petals with white on reverse. Double. B
'Rouletta' (syn. 'Mexicarin' or 'Mexicano'): White petals with variable stripes of red. B
'Snow Queen': Double white with some markings. B
'Wood's Surprise': Green and cream. Pale lilac blooms. E
'Yale': Semi-double. Rich blood red. B

MINIATURE AND DWARF IVY-LEAVED PLANT LIST

'Cascade' types: Of most colours common to the Ivy-leaved range. Smallish flowered and leaved but with many blooms over a long period. Ideal for hanging baskets, etc. B
'Flakey': Variegated green and cream leaves with small off-white blooms with lilac markings. E
'Gay Baby': Small foliage with small off-white blooms and mauve markings. Compact plant. B
'Sugar Baby' (syn. 'Pink Gay Baby'): Dwarf habit, ideal for hanging pots, etc. Double flowers of candy pink. B

HYBRID IVY-LEAVED PLANT LIST

'Blue Spring': Not blue but a bluish mauve, double. E
'Elsi': Cream and green leaves. Bright double orange-scarlet blooms. E
'Millfield Gem': Double shell pink with slight top petal markings. B

ZONAL VARIETIES RECOMMENDED FOR BEDDING PURPOSES

'Caroline Schmidt': Double blooms of bright red with silver variegated foliage.
'Gustav Emich': Scarlet, semi-double.
'Irene Genie': Coral pink semi-double.
'Irene Toyon': Bright red, semi-double.
'King of Denmark': Semi-double, salmon, large heads.
'Paul Crampel': Single vermilion flowers, well-zoned foliage.

PLANTS SUITABLE FOR
BEGINNING A COLLECTION

Pelargonium papillionaceum: A species with large, tri-lobed leaves. Blooms pale mauve with top two petals much larger and marked dark purple.

'Prince of Orange': An orange scented leaf, mid-green in colour. Flowers nice shape and pretty mauve with dark veins.

'Patons Unique': Crimson with neon overlay. Blooms well and for a long period.

'Carisbrooke': A Regal well known for its large pink flowers marked maroon.

'Mrs G.H. Smith': Angel type. Off-white blooms with pink and rosy-mauve markings.

'Doris Moore': Single, zonal. Cherry red with cerise tones.

'Double Henry Jacoby': Deepest red, double flowers.

'Deacon Coral Reef': Good zoned foliage, coral flowers.

'Irene Springtime': Light salmon, semi-double.

'Friesdorfe': A dwarf with small dark green leaves and narrow zone. Single geranium like petals.

'Tom-tit': Unusual miniature with small camellia rose flowers, single.

'Martin Parrett': Miniature with very double blooms of rosy-mauve.

'Mrs J.C. Mappin': Silver variegated foliage with single, off-white flowers and pink tinge.

'Suffolk Gold': Golden foliage type with single red flowers.

'Madam Crousse': Ivy-leaved. Semi-double pink ruffled blooms.

'Mrs W.A.R. Clifton': Double Ivy-leaf with scarlet blooms on vigorous plant.

Appendix I

SUPPLIERS OF PLANTS AND SEEDS

When applying for catalogues or lists please appreciate that there will be a charge for these and for the cost of postage.

Beckwood Geraniums: Beckwood Nurseries Ltd, New Inn Road, Beckley, Oxford OX3 9SS. *Erodium, Geranium, Pelargonium.* Nursery sales and mail order. Catalogue available.

Bressingham Gardens: Blooms of Bressingham Ltd, Diss, Norfolk IP22 2AB. *Erodium, Geranium.* Mail order. Catalogue available.

Brookside: Bedford Road, Holwell, Nr Hitchin, Herts SG5 3RX. *Sarcocaulon, Erodium, Geranium,* succulent *Pelargonium.* Mail order. List available.

Cally Gardens: Gatehouse of Fleet, Castle Douglas, Scotland DG7 2DJ. *Erodium, Geranium.* Mail order. List available.

Chiltern Seeds: Bortree Stile, Ulverston, Cumbria LA12 7PB. Seed of *Erodium, Geranium, Pelargonium.* Mail order. Catalogue available.

Clapton Court Gardens: Crewkerne, Somerset TA18 8PT. *Geranium, Pelargonium.* Nursery sales. Catalogue available.

Coombland Gardens: Coombland, Coneyhurst, Billingshurst, W. Sussex. *Erodium, Geranium.* Nursery sales by apointment only. Mail order. Catalogue available.

Denmead Geranium Nurseries: Hambledon Road, Denmead, Portsmouth PO7 6PS. *Pelargonium.* Mail order. Catalogue available.

155

Ingwersen, W.E.Th.: Birch Farm Nursery, Gravetye, East Grinstead, W. Sussex RH19 4LE. *Erodium, Geranium.* Nursery sales and mail order. Catalogue available.

J. and J. Plants: 16 Bletchingdon Road, Hampton Poyle, Oxford OX5 2QG. *Erodium, Geranium,* Scented-leaved *Pelargonium.* Nursery sales by appointment only. Mail order. List available.

Jumanery Cacti: St Catherine's Lodge, Cranegate Road, Whaplode St Catherine, Nr Spalding, Lincs PE12 6SR. Succulent *Pelargonium.* Mail order. List available.

Kent Street Nurseries: Sedlescombe, Nr. Battle, Sussex. *Pelargonium.* Nursery sales and mail order. List available.

Millbern Geraniums: Pye Court, Willoughby, Rugby CV23 8BZ. *Pelargonium.* Nursery sales and mail order. Catalogue available.

Monica Bennett: Cypress Nursery, Powke Lane, Blackheath, Birmingham. *Pelargonium.* Nursery sales. Catalogue available.

Oakleigh Nurseries: Monkwood, Alresford, Hants SO24 0HB. *Pelargonium.* Nursery sales and mail order. Catalogue available.

Parmenter, F.: 619 Rayleigh Road, Eastwood, Leigh-on-Sea, Essex SS9 5HR. Miniature *Pelargonium* and seed. Nursery sales by appointment.

Riverside Fuchsias and Geraniums: Sutton-at-Hone, Nr Dartford, Kent DA4 9EZ. *Pelargonium.* Nursery sales and mail order. Catalogue available.

Swanland Nurseries (Howard S. Waters and Son): Beech Hill Road, Swanland, Nr Hull HU14 3QY. *Pelargonium.* Nursery sales. Catalogue available.

The Beeches Nursery: 25 Armroyd Lane, Elsecar, Barnsley, S. Yorks S74 8ES. *Pelargonium.* Nursery sales and mail order. List available.

The Red and Green Nurseries: Norwich Road, Dickleburgh, Diss, Norfolk IP21 4NS. *Geranium*. Mail order. Catalogue available.

Thorp's Nurseries: 257 Finchampstead Road, Wokingham, Berks RG11 3JT. *Pelargonium*. Nursery sales and mail order. Catalogue available.

Vernon Geranium Nursery: Cuddington Way, Cheam, Sutton, Surrey SM2 7JB. *Pelargonium*. Mail order. Catalogue available.

Vicarage Garden: Carrington, Urmston, Manchester M31 4AG. *Geranium*. Nursery sales and mail order. Catalogue available.

Appendix   II

SUPPLIERS OF EQUIPMENT AND SUNDRIES

Bunting Biological Control Ltd: The Nurseries, Great Horkesley, Colchester, Essex CO6 4AJ. Biological pest control for the greenhouse and garden. Mail order. Many leaflets available.

Chempak: Geddings Road, Hoddesdon, Herts EN11 0LR. Fertilisers, trace elements, conditioners, etc. Mail order or local nursery. Leaflets available.

Henry Doubleday Research Association: Ryton Gardens, Ryton-on-Dunsmore, Coventry CV8 3LG. Advice, books etc for organic gardening. Mail order. Catalogue available.

Mountain Breeze Air Ionizers: Peel Road, Skelmersdale, Lancs WN8 0PT. Mail order. Advice leaflet available.

Phostrogen Ltd: Corwen, Clwyd LL21 0EE. Fertilisers and other sundries. Mail order or local nursery. Many leaflets on plant care available.

Rapitest: Wilson Grimes Products, London Road, Corwen, Clwyd. Meters for soil, watering and light testing. Local nursery. Leaflets available.

Silverperl Products Ltd: Sales Office, Dept H 301, P O Box 8, Harrogate, N. Yorks HG2 8JW. Perlite, vermiculite, etc. Leaflets on plant care available.

Appendix III

NATIONAL SOCIETIES

Many specialist societies cater for the Geraniaceae enthusiast in one form or another. They require an annual fee or subscription and because these cannot remain static it is advised that the societies be approached individually for up-to-date advice and information.

British Cactus and Succulent Society: Hon. Sec. Mr J.W.P. Mullard, 19 Crabtree Road, Botley, Oxford OX2 9DU. *Sarcocaulon* and the succulent *Pelargonium* are catered for in their magazines and show schedules. There are many local branches in Great Britain.

British and European Geranium Society: Hon. Sec. Mr A. Biggin, 'Morval', The Hills, Bradwell, Sheffield S30 2HZ. The membership is mainly for the *Pelargonium* grower but deals with the whole family. Meetings, shows, conferences and quarterley publications are offered; also regional groups as well as members in other countries.

British Pelargonium and Geranium Society: Hon. Sec. Mrs Jan Taylor, 23 Beech Crescent, Kidlington, Oxford OX5 1DW. The society promotes interest in all members of the Geraniaceae family. Four regular magazines, shows, meetings, book sales and advice leaflets are organised. Free seeds for members most years. Membership is worldwide. There are local representatives and a Geraniaceae group which deals in detail with the species of the family and carries out a seed redistribution scheme.

Hardy Plant Society: Mrs J. Sterndale-Bennett, White Windows, Longparish, Hants SP11 6PB. Formed to create an interest in hardy perennial plants it now boasts a *Geranium* group. Newsletters and society magazines are sent throughout the year. A seed redistribution scheme operates.

Henry Doubleday Research Association: Address in suppliers section. Their aims are to promote organic gardening and crop production. A quarterly newsletter, meetings, etc. are offered to members.

National Council for the Conservation of Plants and Gardens: Gen. Sec. Mr R.A.W. Lowe, Wisley Gardens, Woking, Surrey GU23 6QB. Among the objects are to encourage the conservation of endangered or rare plants. Meetings, displays, shows together with the establishment of national collections of plants throughout Great Britain are the many benefits to members. Local groups are in existence.

The Royal Horticultural Society: The Secretary, 80 Vincent Square, London SW1P 2PE.

Appendix IV

INTERNATIONAL SOCIETIES AND GROUPS

There is a good deal of interest in the Geraniaceae family in many countries – the promotion and breeding of cultivars and also the conservation of species is widespread. It is a close-knit Fancy or Hobby and it is only a matter of months before news of new cultivars or other major events is shared. There are continual exchanges of advice, information and the new cultivars soon find their way into catalogues everywhere. Many countries abroad possess their own societies:

Australian Geranium Society Inc.: Sec. Mrs Grace Perry, 118 Thornley Road, Fairfield West, New South Wales 2165, Australia. The aims are to promote the culture and knowledge of the Geraniaceae in all its phases.

Nederlandse Pelargonium en Geranium Vereniging: R. van der Lee, Hollands End 89, 1244 NP Ankeveen, Netherlands. Their aim and activities are to popularise *Pelargonium* and *Geranium* for a larger public by organising shows, exhibitions and markets.

International Geranium Society: Mrs Robin Schultz, 4610 Druid Street, Los Angeles, California 90032, USA. A quarterly magazine called *Geraniums around the World*, a seed exchange system and advice on *Pelargonium*.

South African Geranium and Pelargonium Society: P.O. Box 8714, Johannesburg 2000, South Africa. A four times a year magazine, meetings, trials, shows, field days and a library are offered to members.

South Australia Geranium and Pelargonium Society: Hon. Sec. Miss P.M. Lamphee, 244 Young Street, Unley 5061, Adelaide, Australia. Their aims are to promote the knowledge and growth of the plants by holding meetings and shows.

Victoria (Australia) Pelargonium and Geranium Society: Sec. Mrs Lila Rechter, 46 Edmonds Avenue, Ashwood 3147, Australia. To improve and encourage the study of the family and provide facilities for identification of cultivars and species are the main objects.

Western Australian Geranium and Pelargonium Society Inc.: Hon. Sec. Mrs J. Tondut, 42 Kathleen Street, North Cottesloe 6011, Western Australia. Their aim is to foster the love of geraniums and pelargoniums throughout the state by holding meetings, shows and displays.

Appendix V

GLOSSARY

Acaulescent   With short stem hardly noticeable.
Actinomorphic   Regular, in shape, flower formation.
Annual   Growing to maturity in one year or less.
Anther   The pollen bearing part of the stamen.
Bract   Modified leaves on the inflorescence that protect, as a covering, the flower head in bud.
Calyx   The outer flower part used to protect the single flower.
Capitate   Growing in a head.
Capsule   Generally, the seed container.
Carpel   The part in which the seeds are formed.
Chlorophyll   The green pigment in plants essential for survival. It is manufactured by the process of the plant absorbing light and using the energy in the process of photosynthesis to build sugars from water and carbon dioxide.
Chlorosis   The poor production of chlorophyll.
Cordate   Heart-shaped.
Corymb   Flat topped group of flowers.
Cotyledon   The first seedling leaf to appear.
Crenate   Rounded teeth at leaf margin.
Cultivar   A cultivated variety.
Cuneate   Wedge-shaped.
Dentate   Blunt teeth at leaf margin.
Entire   Smooth continuous edges to the leaf margin.
Fertile   Capable of producing seed.
Fertilisation   The fusing of male and female sex cells.
Filament   The anther's stalk.
Genus (plural *genera*)   Botanical category of the classification of a group of plants with common characteristics evolved from, probably, a common ancestor and agreed by botanists.
Geophyte   A plant having an underground tuber for storing food.
Glabrous   Hairless.
Glaucous   With bluish tint caused by a bloom or waxy coating (patina), e.g. grapes.

165

Habitat   Home of plants in the wild.

Hybrid   A plant resulting from the crossing of two distinct types or species or sometimes genera.

Inflorescence   Arrangement of flowers, e.g. spike, lavender; umbel, allium.

Lanceolate   Shaped like the head of a lance, i.e. broad at the base, tapering to a point at the tip and at least three times as long as the width.

Lobed   A part of leaf or flower divided from the rest of the part but not totally.

Mericarp   One of the five-seeded portions.

Node   The place or joint where the leaf stalks meet the stem.

Palmate   Hand-shaped, describing a leaf as an open hand.

Pedicel   Stalk of a single flower.

Peduncle   The stem supporting the flower arrangement.

Peltate   Joined to the middle.

Perennial   With a life-span of some years.

Petiole   Stalk of a leaf.

Pinnatifid   Divided half way to the mid-rib.

Pinnatipartite   Separate lobes of feathered leaf.

Pistil   The female organ comprising an ovary, stigma and style (rostrum).

Pubescent   With short hairs.

Reniform   Kidney-shaped.

Rhizome   Horizontal underground stem used for storage.

Rib   One of the main prominent leaf veins.

Rostrum   see *Style*.

Serrate   Saw-like teeth.

Species   A group of closely allied plants within a genus, having essential characteristics that are distinctive and consistently breed true to type from seed for generations.

Sport   A shoot differing from the host plant as a result of spontaneous change in its hereditary genes. In the wild it could be the result of cosmic radiation. It can be simulated by man with the use of drugs and X-ray processes.

Stamen   The male element of a flower.

Stigma   The female element of a flower.

Stipule   Scales in pairs on the stem at the base of the petioles.

Stolon   Prostrate stem rooting at the nodes.

Style   The stalk linking the ovary and the stigma.

Sub-shrub   Shrub-like plant where only the base is woody.

Succulent   Fleshy.

Tomentose   Covered with hairs.

Tripartite   Divided nearly to the base in three segments.

Tuber   Thickened fleshy root or underground stem for food storage.

Umbel   An inflorescence with all pedicels coming from the top of the peduncle.

Xerophytic   A plant capable of surviving long dry periods either by the formation of modified spines or dense hairs etc.

Zygomorphic   Irregular flower capable of being divided into equal halves in one plane only.

Appendix VI

FURTHER READING

The Pelargonium Family, William J. Webb, Croom Helm (1984)

Hardy Geraniums, Peter Yeo, Croom Helm (1985)

Geraniums for Home and Garden, Alan Shellard, David & Charles (1981)

Growing and Showing Geraniums, Alan Shellard, David & Charles (1984)

International Pelargoniums or Geraniums of the World, Henry J. Wood, Henry J. Wood (1983)

Pelargoniums, Geraniums and Their Societies, Henry J. Wood, Henry J. Wood (1974)

Pelargoniums: A Complete Guide to their Cultivation, Henry J. Wood, Henry J. Wood (1966)

Pelargoniums: The Grower's Guide to 'Geraniums' especially the Golden-Leaved Varieties, Henry J. Wood, Henry J. Wood (1987)

Geraniums and Pelargoniums, H. G. Witham Fogg, John Gifford Ltd (1975)

General Index

General Index

Index of Species
and Cultivars

175